ESTATE PUBLICATIONS

IEW FOREST

HRISTCHURCH · RINGWOOD

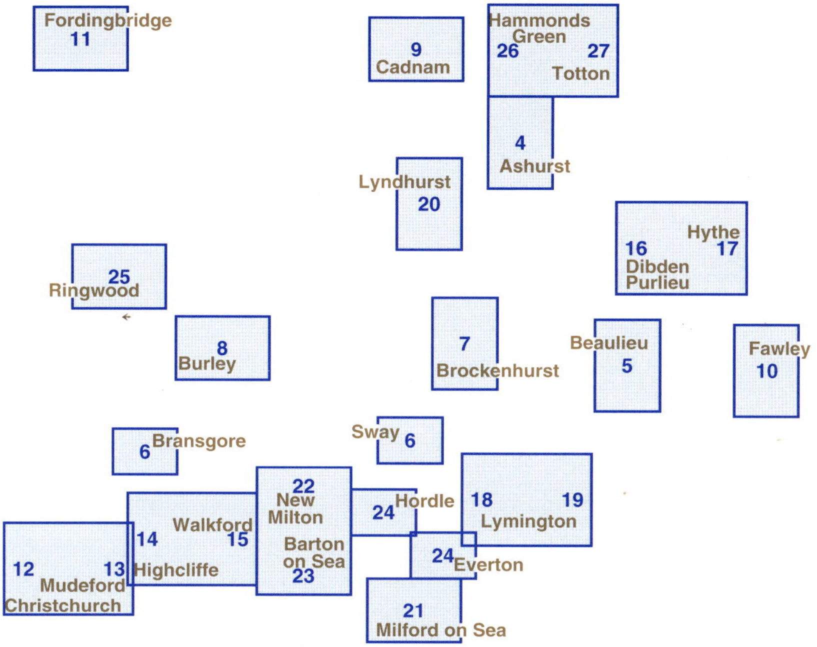
Fordingbridge
11
9
Cadnam
Hammonds Green
26
27
Totton
4
Ashurst
Lyndhurst
20
Hythe
16
17
Dibden Purlieu
25
Ringwood
8
Burley
7
Brockenhurst
Beaulieu
5
Fawley
10
6
Bransgore
Sway
6
22
New Milton
24
Hordle
18
19
Lymington
14
Walkford
15
12
13
Highcliffe
Mudeford
Christchurch
Barton on Sea
23
24
Everton
21
Milford on Sea

OAD MAP Page 2
DEX TO STREETS Page 28

Scale of street plans: 4 Inches to 1 Mile (unless otherwise stated)

- Motorway
- 'A' Road / Dual
- 'B' Road / Dual
- Minor Road / Dual
- Track
- Pedestrianized
- Railway / Station
- Footpath

Every effort has been made to verify the accuracy of information in this book but the publishers cannot accept responsibility for expense or loss caused by an error or omission. Information that will be of assistance to the user of the maps will be welcomed.

The representation on these maps of a road, track or path is no evidence of the existence of a right of way.

- Stream / River
- Lock Canal
- One-way Street
- Car Park
- Public Convenience
- Tourist Information
- Place of Worship
- Post Office

treet plans prepared and published by ESTATE PUBLICATIONS, Bridewell House, TENTERDEN, KENT.
The Publishers acknowledge the co-operation of the local authorities of towns represented in this atlas.

Ordnance Survey® This product includes mapping data licensed from Ordnance Survey® with the permission of the Controller of Her Majesty's Stationery Office.

Estate Publications 386 E

ISBN 1 84192 113 0

ROAD MAP

Scale: 3 Miles to 1 Inch

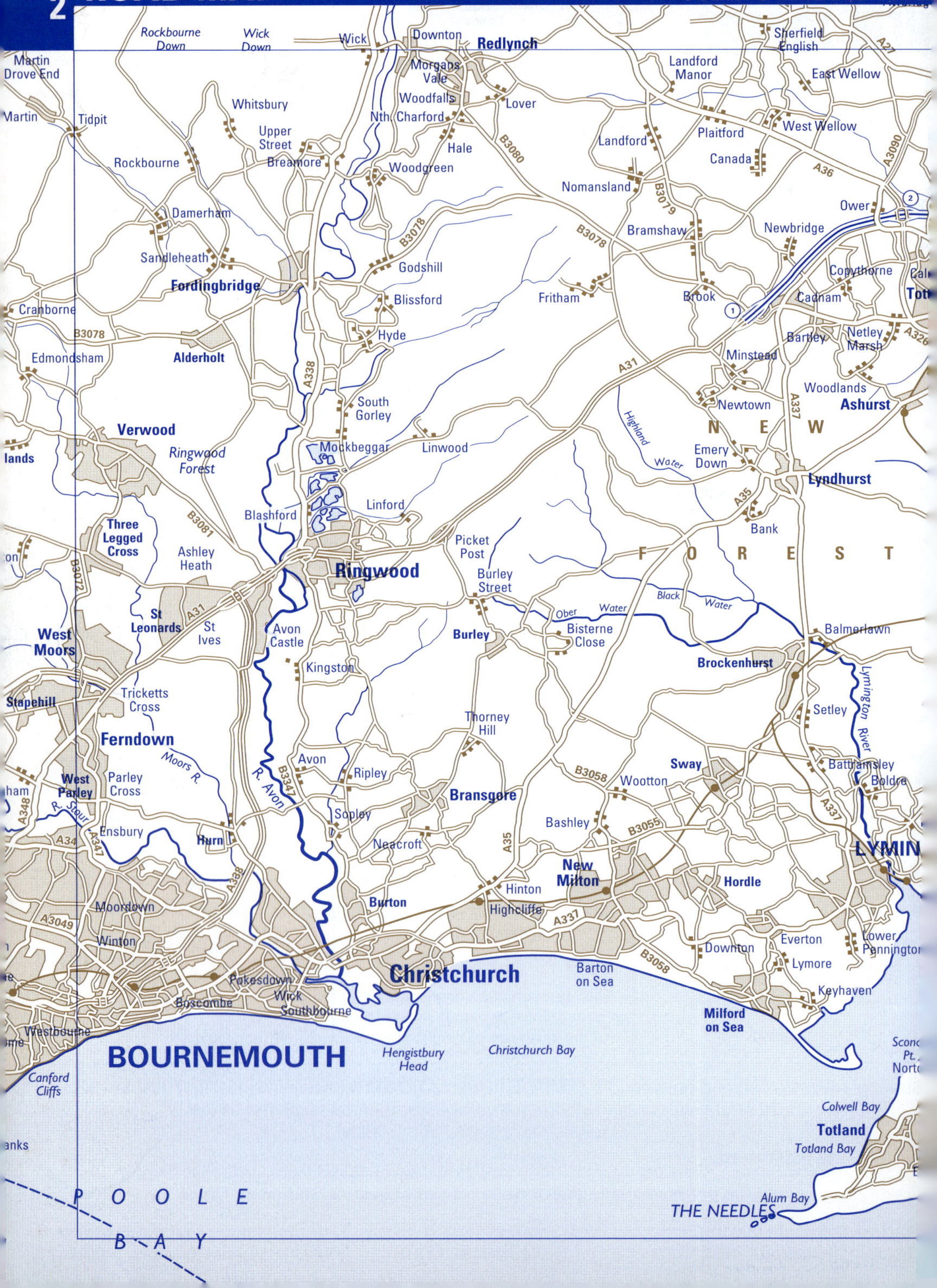
Rockbourne Down
Wick Down
Wick
Downton
Redlynch
Sherfield English
A27
Martin Drove End
Morgans Vale
Landford Manor
East Wellow
Whitsbury
Woodfalls
Lover
Martin
Tidpit
Nth Charford
West Wellow
Upper Street
Plaitford
Landford
Hale
B3080
A3090
Rockbourne
Breamore
Canada
Woodgreen
A36
Nomansland
B3079
Ower
Damerham
Bramshaw
Newbridge
B3078
Sandleheath
Godshill
Fordingbridge
Copythorne
Blissford
Fritham
Brook
Cadnam
Cranborne
B3078
Hyde
Bartley
Netley Marsh
A326
Alderholt
Edmondsham
Minstead
A31
A338
Woodlands
South Gorley
Newtown
Ashurst
A337
Verwood
Highland Water
N E W
lands
Mockbeggar
Linwood
Emery Down
Ringwood Forest
Lyndhurst
A35
Linford
Blashford
B3081
Bank
Three Legged Cross
Picket Post
Ashley Heath
F O R E S T
Ringwood
B3072
Burley Street
Black Water
St Leonards
A31
Ober Water
St Ives
West Moors
Avon Castle
Burley
Bisterne Close
Balmerlawn
Brockenhurst
Kingston
Lymington River
Stapehill
Tricketts Cross
Setley
Thorney Hill
Ferndown
Moors R.
Avon
R. Avon
Sway
B3347
B3058
Ripley
Wootton
Battramsley
Boldre
West Parley
Parley Cross
Bransgore
A348
R. Stour
Sopley
Bashley
A337
Ensbury
B3055
A34
A347
Hurn
Neacroft
A35
New Milton
LYMIN
A338
Hinton
Hordle
Burton
Moordown
Highcliffe
A3049
A337
Winton
Downton
Everton
Lower Pennington
Lymore
Pokesdown
Christchurch
Barton on Sea
B3058
Wick
Keyhaven
Boscombe
Southbourne
Milford on Sea
Westbourne
BOURNEMOUTH
Hengistbury Head
Christchurch Bay
Canford Cliffs
Colwell Bay
Totland
Totland Bay
Alum Bay
THE NEEDLES
P O O L E
B A Y

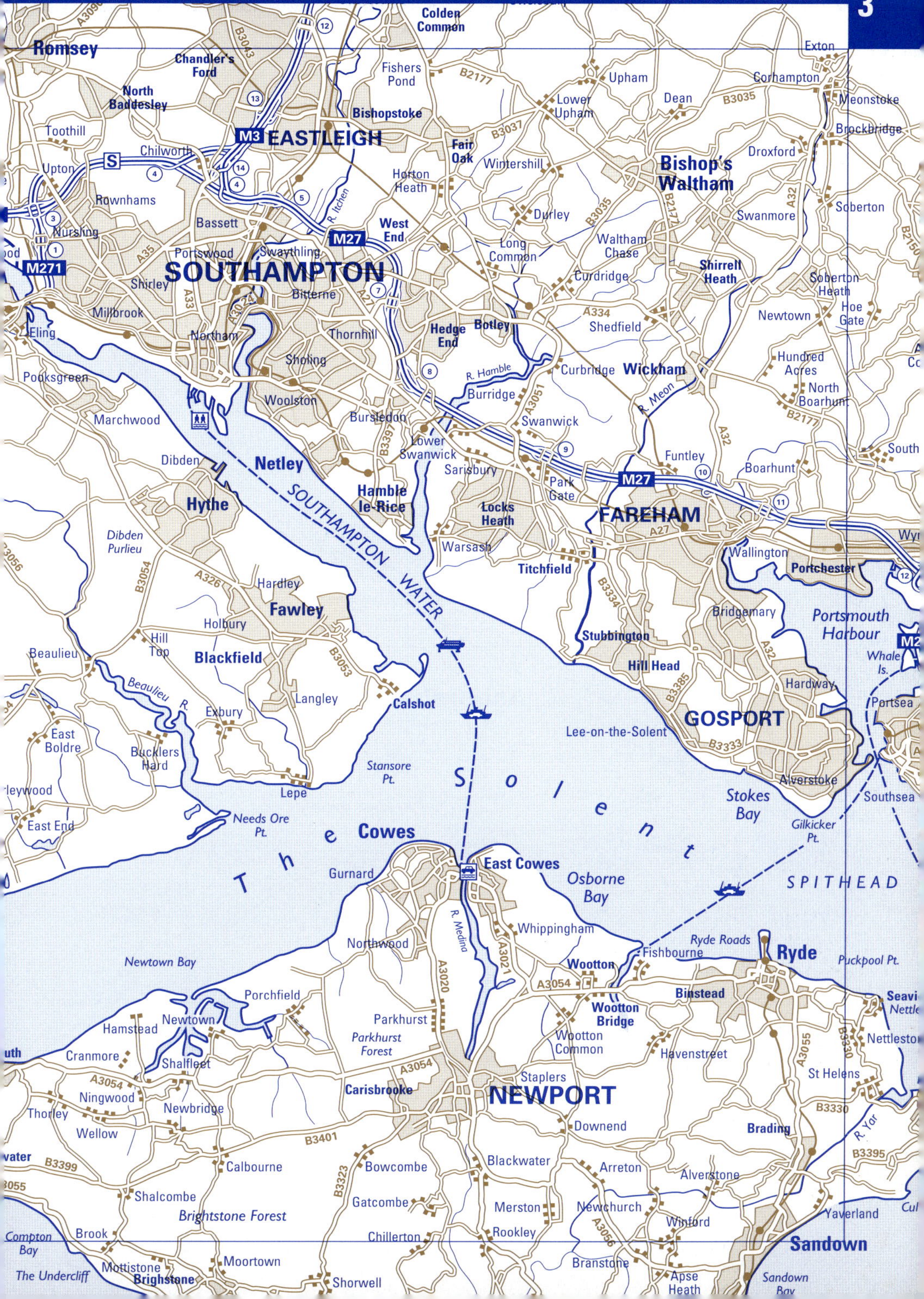
Romsey
Chandler's Ford
Colden Common
Fishers Pond
B2177
Upham
Exton
Corhampton
Meonstoke
Lower Upham
Dean
B3035
North Baddesley
Bishopstoke
M3
EASTLEIGH
Toothill
Chilworth
Fair Oak
B3037
Wintershill
Brockbridge
Droxford
Bishop's Waltham
Upton
S
Horton Heath
Rownhams
Nursling
Bassett
R. Itchen
West End
M27
Durley
B3035
Swanmore
A32
Soberton
M271
Portswood
Swaythling
SOUTHAMPTON
Long Common
Waltham Chase
Shirrell Heath
Shirley
A35
A33
Bitterne
A27
Curdridge
Soberton Heath
Hoe Gate
Millbrook
Eling
Northam
Thornhill
Hedge End
Botley
A334
Shedfield
Newtown
Sholing
R. Hamble
Curbridge
Wickham
Hundred Acres
Pooksgreen
North Boarhunt
R. Meon
Burridge
A3051
Woolston
B2177
Marchwood
Bursledon
Swanwick
Lower Swanwick
B3397
Sarisbury
Funtley
South
Dibden
Netley
A32
Boarhunt
Park Gate
Hythe
SOUTHAMPTON WATER
Hamble le-Rice
Locks Heath
FAREHAM
A27
Dibden Purlieu
Warsash
Titchfield
Wallington
Portchester
B3054
A326
Hardley
Fawley
B3334
Bridgemary
Portsmouth Harbour
Holbury
Hill Top
Stubbington
Whale Is.
Beaulieu
Blackfield
B3053
Hill Head
A32
Beaulieu R.
Hardway
Langley
B3385
Exbury
Calshot
Portsea
East Boldre
Lee-on-the-Solent
GOSPORT
B3333
Bucklers Hard
Stansore Pt.
Alverstoke
Lepe
The Solent
Stokes Bay
Southsea
East End
Needs Ore Pt.
Cowes
Gilkicker Pt.
East Cowes
Osborne Bay
SPITHEAD
Gurnard
R. Medina
Whippingham
Northwood
A3021
Ryde Roads
Fishbourne
Ryde
Puckpool Pt.
Newtown Bay
A3020
Wootton
A3054
Binstead
Porchfield
Wootton Bridge
Parkhurst
Newtown
Hamstead
Parkhurst Forest
Wootton Common
Havenstreet
A3055
B3330
Nettlestone
Cranmore
Shalfleet
A3054
St Helens
Staplers
A3054
Ningwood
Carisbrooke
NEWPORT
Thorley
Newbridge
B3330
Wellow
Downend
Brading
R. Yar
B3401
B3399
Calbourne
Bowcombe
Blackwater
Arreton
B3395
B3323
Alverstone
Shalcombe
Gatcombe
Merston
Newchurch
Brightstone Forest
A3056
Winford
Yaverland
Brook
Compton Bay
Chillerton
Rookley
Sandown
Moortown
Branstone
Mottistone
The Undercliff
Brighstone
Shorwell
Apse Heath
Sandown Bay

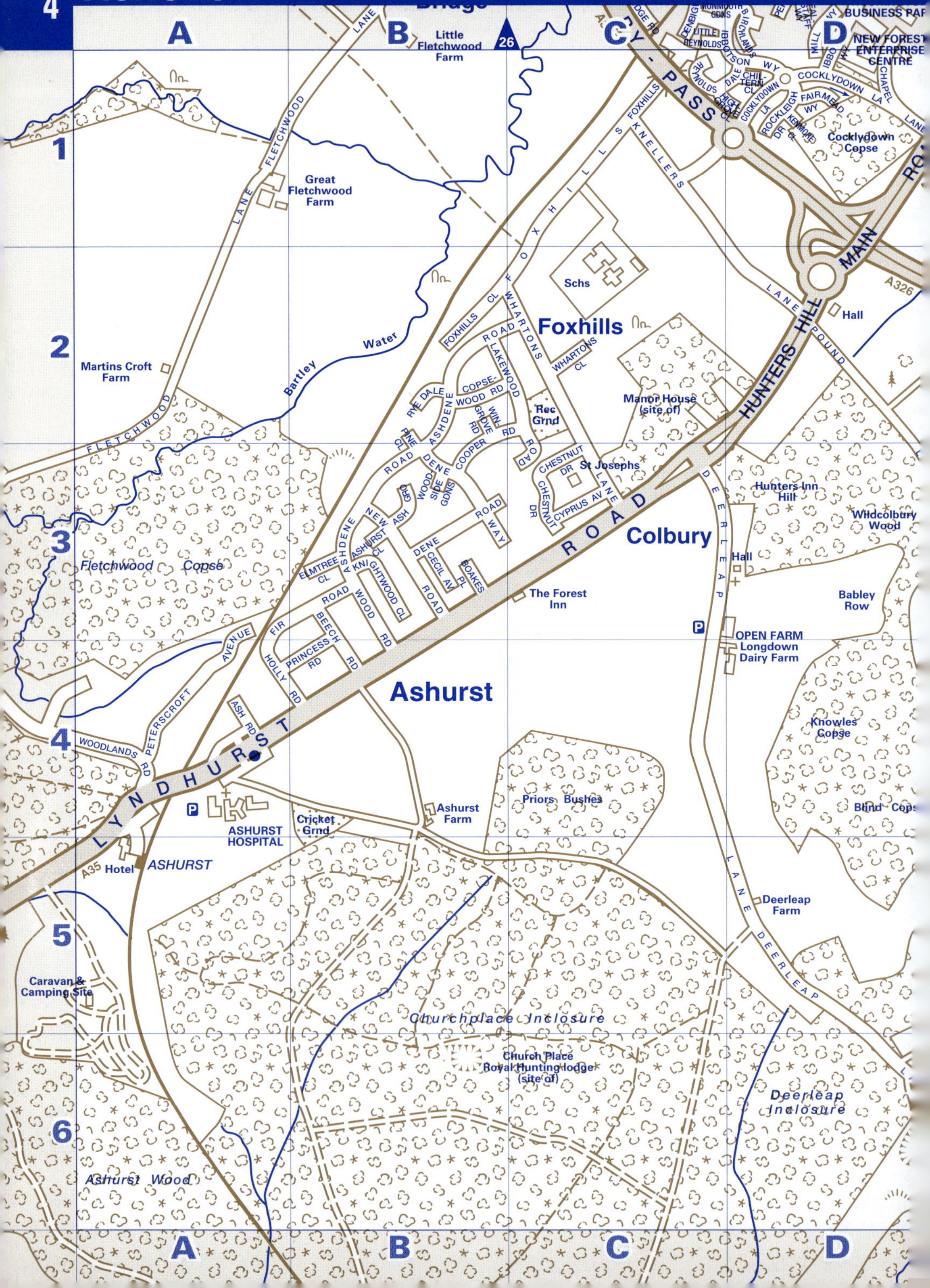
A
B
C
D
1
2
3
4
5
6
26
Little Fletchwood Farm
Great Fletchwood Farm
FLETCHWOOD LANE
Martins Croft Farm
Bartley Water
Foxhills
Schs
Rec Grnd
Manor House (site of)
St Josephs
Colbury
Hunters Inn Hill
Wildcolbury Wood
Hall
Babley Row
OPEN FARM
Longdown Dairy Farm
Fletchwood Copse
The Forest Inn
Ashurst
Knowles Copse
Priors Bushes
Ashurst Farm
Cricket Grnd
ASHURST HOSPITAL
Hotel
ASHURST
A35
LYNDHURST ROAD
Blind Copse
Deerleap Farm
DEERLEAP LANE
Caravan & Camping Site
Churchplace Inclosure
Church Place Royal Hunting lodge (site of)
Deerleap Inclosure
Ashurst Wood
Cocklydown Copse
NEW FOREST ENTERPRISE CENTRE
BUSINESS PAR
HUNTERS HILL
MAIN
A326
POUND LANE
WOODLANDS RD
PETERSCROFT AVENUE
FOXHILLS
KNELLERS
WHARTONS
Hall

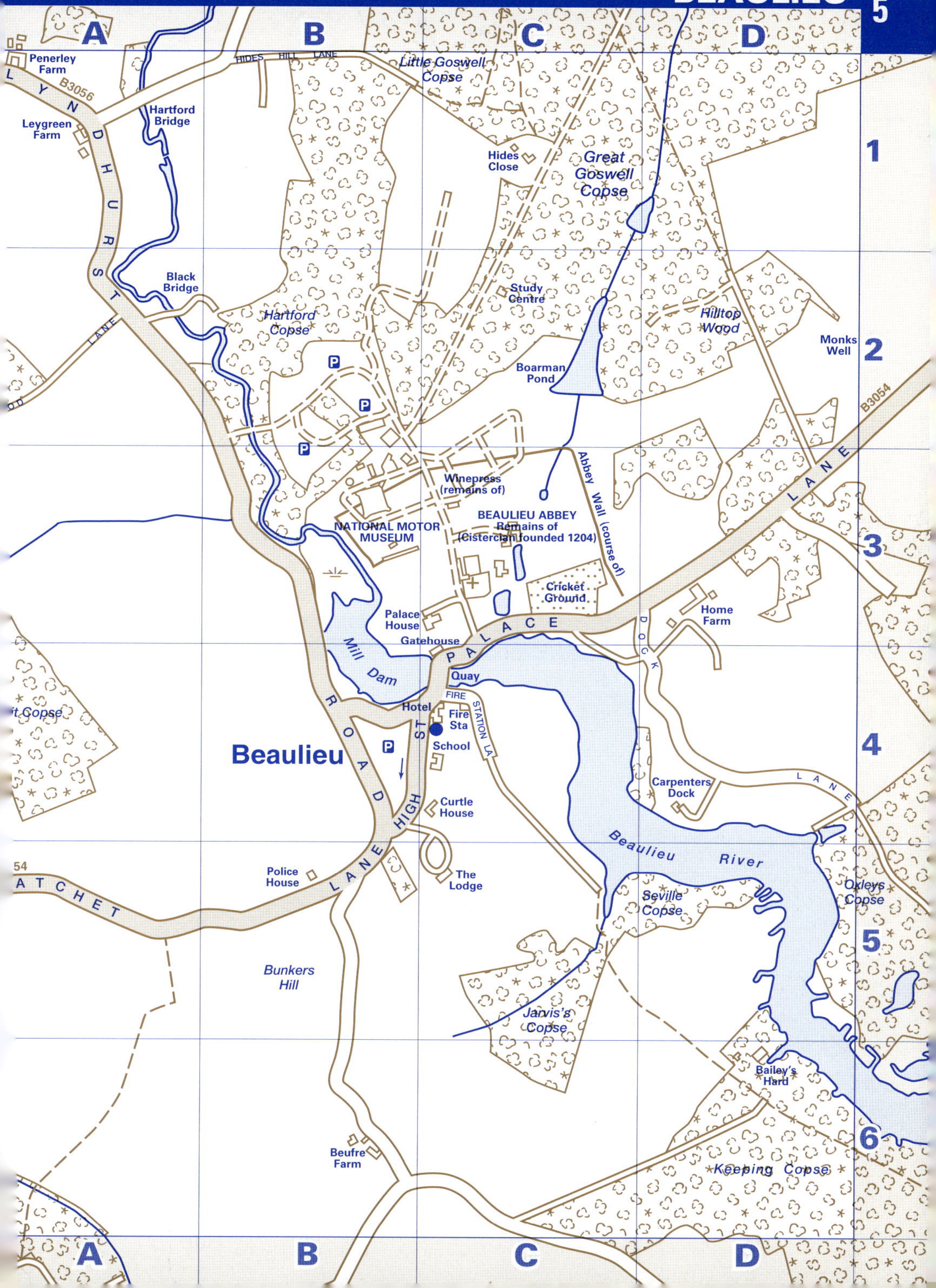
A
B
C
D
1
2
3
4
5
6
Penerley Farm
LYNDHURST
B3056
HIDES HILL LANE
Little Goswell Copse
Leygreen Farm
Hartford Bridge
Hides Close
Great Goswell Copse
Black Bridge
Study Centre
Hartford Copse
Hilltop Wood
LANE
Monks Well
Boarman Pond
B3054
P
Abbey Wall (course of)
Winepress (remains of)
BEAULIEU ABBEY Remains of (Cistercian founded 1204)
NATIONAL MOTOR MUSEUM
LANE
Cricket Ground
Palace House
PALACE
Home Farm
Gatehouse
DOCK
Mill Dam
Quay
FIRE STATION LA
Hotel
Fire Sta
ROAD
ST
School
Beaulieu
Carpenters Dock
LANE
HIGH
Curtle House
Beaulieu River
54
Police House
The Lodge
Oxleys Copse
HATCHET LANE
Seville Copse
Bunkers Hill
Jarvis's Copse
Bailey's Hard
Beufre Farm
Keeping Copse

BRANSGORE

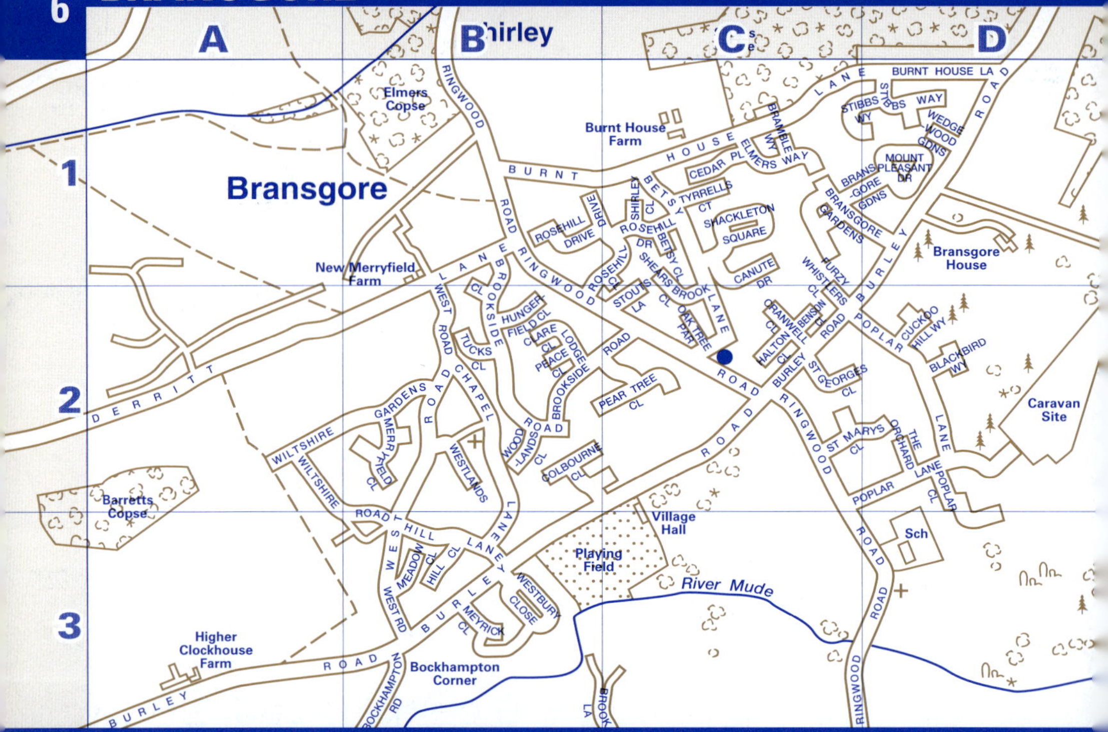
A
B
C
D
1
2
3
Shirley
Elmers Copse
Bransgore
Burnt House Farm
BURNT HOUSE LANE
BURNT HOUSE LA
RINGWOOD ROAD
STIBBS WY
STIBBS WAY
WEDGE WOOD GDNS
MOUNT PLEASANT DR
BRAMBLE WY
ELMERS WAY
CEDAR PL
TYRRELLS CT
SHACKLETON SQUARE
BRANSGORE GDNS
BRANSGORE GARDENS
BURLEY ROAD
Bransgore House
ROSEHILL DRIVE
ROSEHILL CL
ROSEHILL DR
SHIRLEY CL
BETSY CL
BETSY LANE
SHEARS BROOK
STOUTS LA
OAK TREE PAR
CANUTE DR
WHISTLERS CL
FURZY CL
CRANWELL CL
HALTON CL
BENSON ROAD
POPLAR LANE
CUCKOO HILL WY
BLACKBIRD WY
Caravan Site
New Merryfield Farm
DERRITT LANE
WEST ROAD
BROOKSIDE CL
BROOKSIDE ROAD
HUNGERFIELD CL
TUCKS CL
CLARE LODGE CL
PEACE CL
PEAR TREE CL
BURLEY CL
ST GEORGES CL
CHAPEL LANE
GARDENS ROAD
MERRYFIELD CL
WILTSHIRE ROAD
WILTSHIRE
WESTLANDS
WOODLANDS ROAD
COLBOURNE CL
ST MARYS CL
THE ORCHARD
POPLAR
Barretts Copse
Village Hall
Playing Field
Sch
WEST HILL ROAD
MEADOW CL
HILL CL
WESTBURY CLOSE
MEYRICK CL
WEST RD
River Mude
Higher Clockhouse Farm
BURLEY ROAD
Bockhampton Corner
BOCKHAMPTON RD
BROOK LA

SWAY

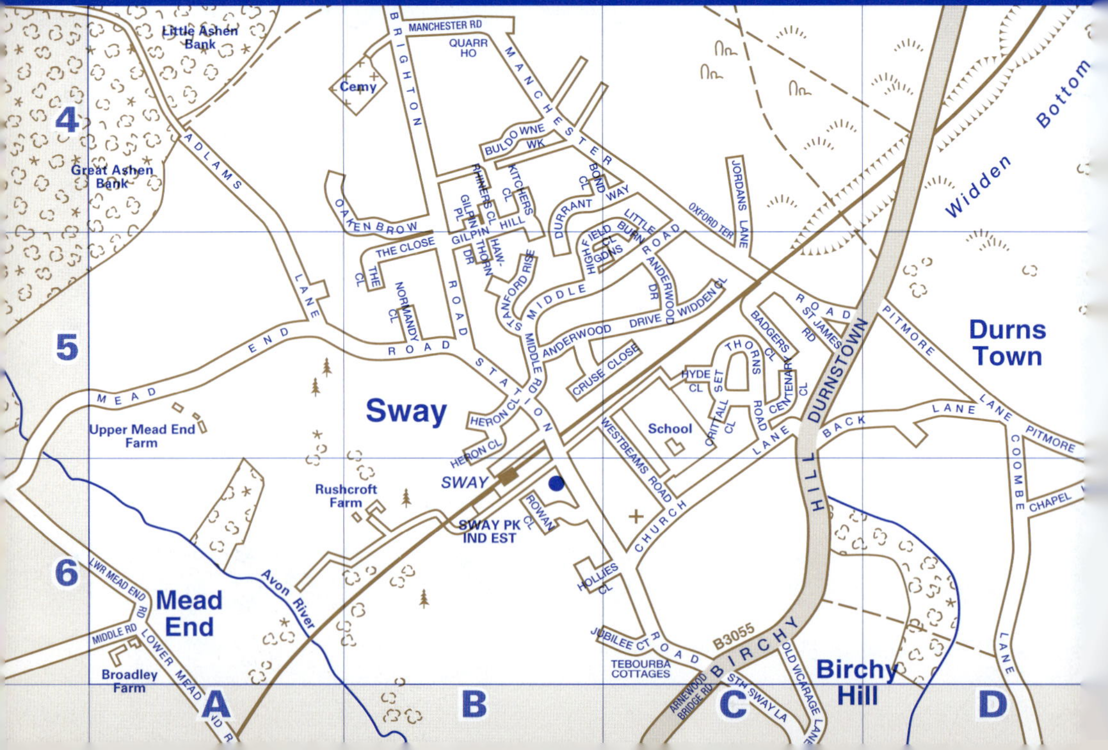
4
5
6
A
B
C
D
Little Ashen Bank
Great Ashen Bank
Cemy
MANCHESTER RD
QUARR HO
BRIGHTON ROAD
MANCHESTER ROAD
BULDOWNE WK
KITCHERS CL
RHINEFIELD CL
GILPIN PL
GILPIN HILL
GILPIN DR
HAWTHORN DR
BOND CL
DURRANT WAY
LITTLE BURN
HIGHFIELD GDNS
OXFORD TER
JORDANS LANE
ADLAMS LANE
OAKEN BROW
THE CLOSE
THE CL
NORMANDY CL
STANFORD RISE
MIDDLE ROAD
MIDDLE RD
ANDERWOOD DR
ANDERWOOD DRIVE
WIDDEN CL
ST JAMES RD
BADGERS CL
THORNS LANE
SET
HYDE CL
CRITTALL CL
CENTENARY CL
Widden Bottom
Durns Town
PITMORE ROAD
MEAD END ROAD
Upper Mead End Farm
Sway
HERON CL
STATION ROAD
CRUSE CLOSE
School
DURNSTOWN
BACK LANE
PITMORE LANE
COOMBE LANE
CHAPEL
SWAY
SWAY PK IND EST
ROWAN CL
WESTBEAMS ROAD
CHURCH LANE
Rushcroft Farm
Avon River
LWR MEAD END RD
Mead End
MIDDLE RD
LOWER MEAD RD
Broadley Farm
HOLLIES CL
JUBILEE CT
TEBOURBA COTTAGES
B3055
BIRCHY HILL
ARNEWOOD BRIDGE RD
STH SWAY LA
OLD VICARAGE LANE
Birchy Hill
LANE

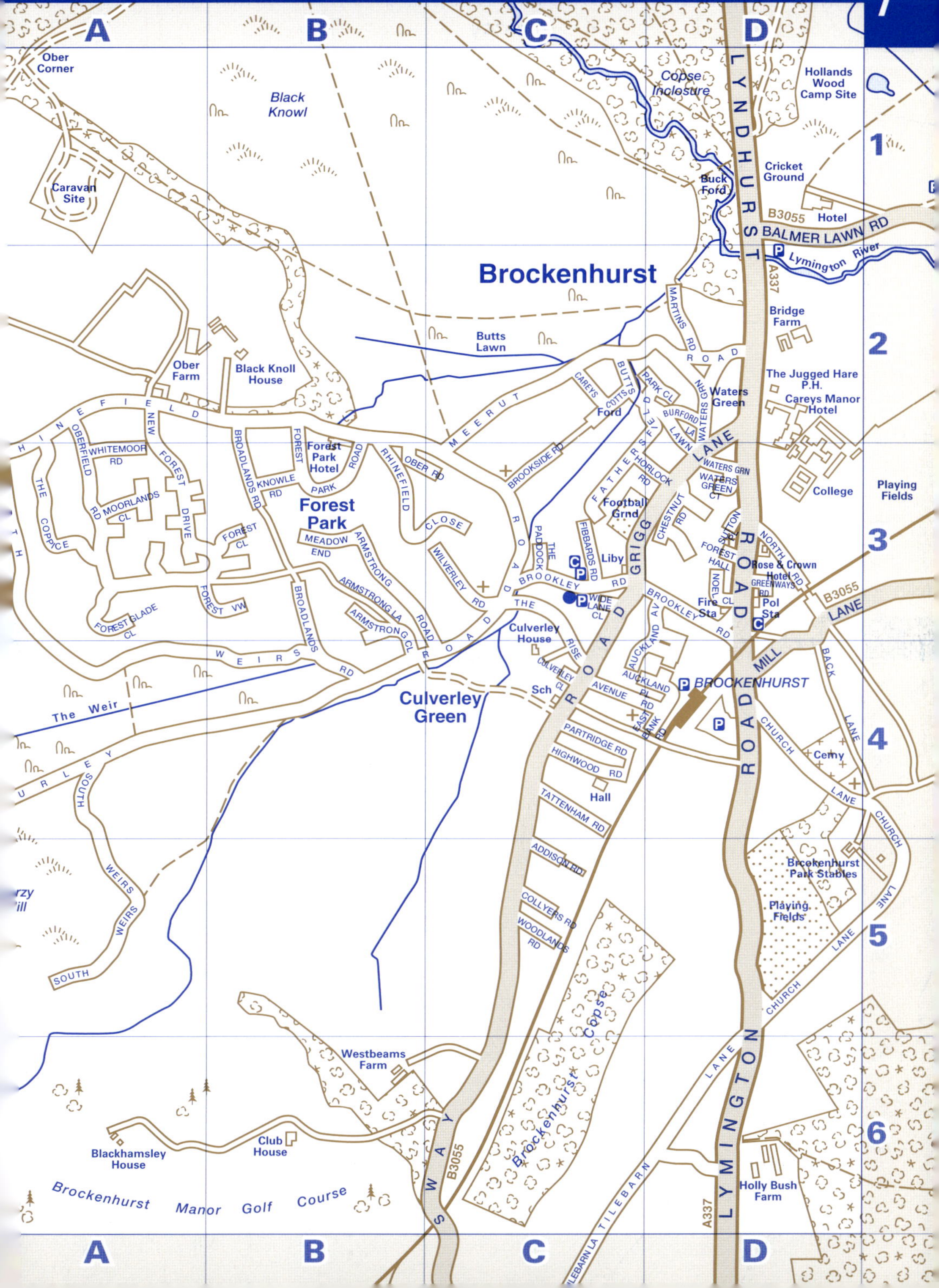

Brockenhurst
Ober Corner
Black Knowl
Caravan Site
Copse Inclosure
Hollands Wood Camp Site
Cricket Ground
Buck Ford
Hotel
B3055
BALMER LAWN RD
Lymington River
A337
LYNDHURST ROAD
Bridge Farm
MARTINS RD
ROAD
Butts Lawn
Ober Farm
Black Knoll House
The Jugged Hare P.H.
Careys Manor Hotel
Waters Green
HINEFIELD RD
MEERUT RD
CAREYS COTTS
Ford
BUTTS
PARK CL
BURFORD LA
WATERS GRN
LAWN LANE
WHITEMOOR RD
OBERFIELD RD
NEW FOREST DRIVE
BROADLANDS RD
FOREST RD
Forest Park Hotel
RHINEFIELD RD
OBER RD
BROOKSIDE RD
FATHERSFIELD
HORLOCK RD
WATERS GRN
WATERS GREEN CT
College
Playing Fields
THE COPPICE
MOORLANDS CL
KNOWLE RD
PARK
Forest Park
CLOSE
Football Grnd
CHESTNUT RD
SUTTON RD
NORTH RD
FOREST CL
MEADOW END
ARMSTRONG RD
WILVERLEY RD
THE PADDOCK
FIBBARDS RD
Liby
GRIGG
FOREST HALL
Rose & Crown Hotel
GREENWAYS RD
B3055
LANE
BROOKLEY RD
WIDE LANE CL
NOEL CL
Fire Sta
Pol Sta
FOREST VW
ARMSTRONG LA
ARMSTRONG CL
THE RISE
BROOKLEY RD
AUCKLAND AV
FOREST GLADE CL
BROADLANDS
WEIRS RD
Culverley House
CULVERLEY CL
AUCKLAND PL
MILL
BACK LANE
Sch
AVENUE RD
BROCKENHURST
EAST BANK RD
The Weir
Culverley Green
ROAD
PARTRIDGE RD
CHURCH LANE
Cemy
RHINEFIELD
HIGHWOOD RD
SOUTH WEIRS
Hall
TATTENHAM RD
CHURCH LANE
ADDISON RD
Brockenhurst Park Stables
Playing Fields
COLLYERS RD
WOODLANDS RD
Copse
CHURCH LANE
Westbeams Farm
Brockenhurst Copse
LYMINGTON ROAD
LANE
SWAY
B3055
TILEBARN LA
Blackhamsley House
Club House
Holly Bush Farm
Brockenhurst Manor Golf Course
A337
A B C D
1 2 3 4 5 6

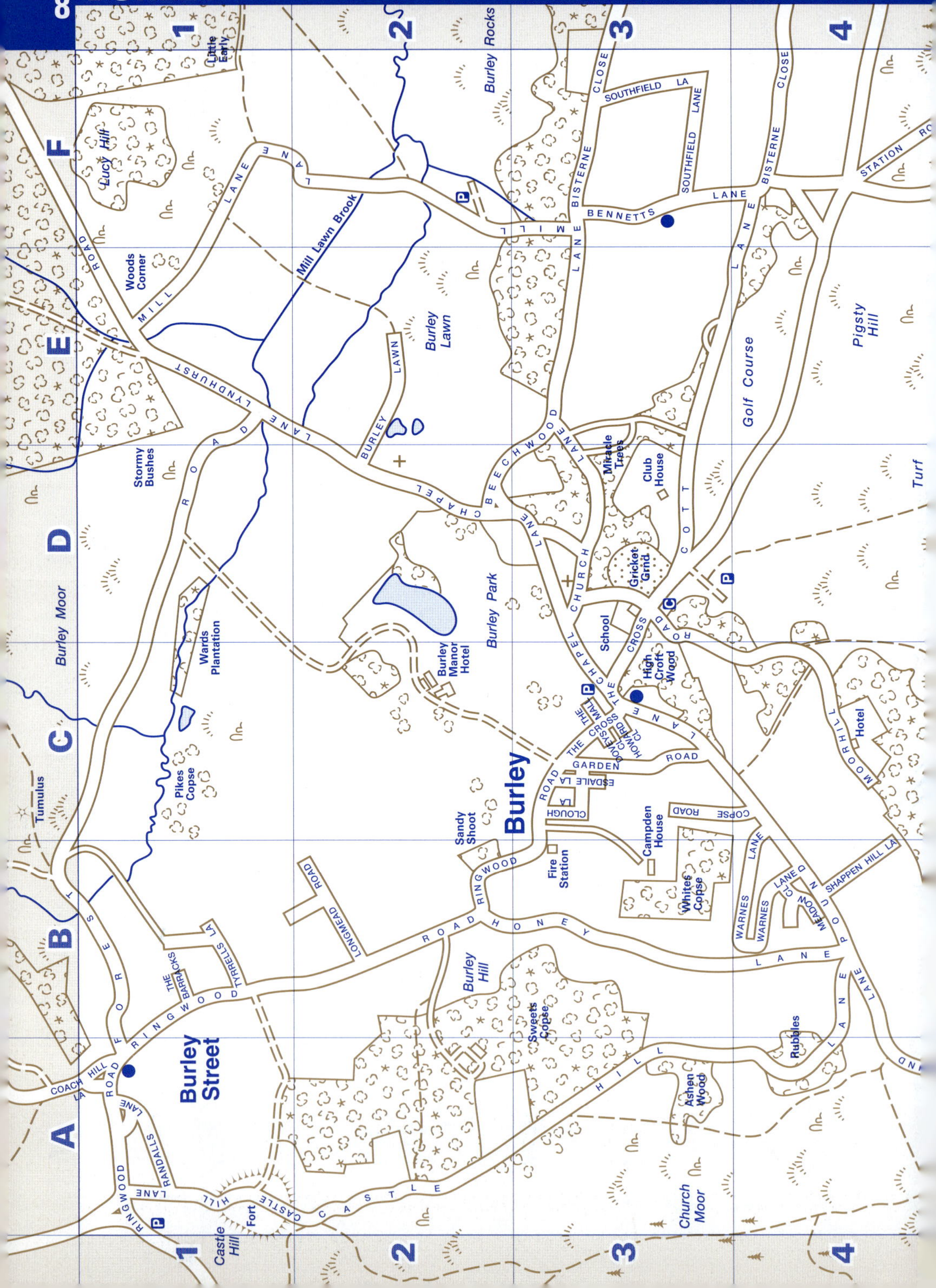
Burley
Burley Street
Burley Moor
Burley Rocks
Burley Lawn
Burley Park
Burley Hill
Burley Manor Hotel
Little Early
Lucy Hill
Woods Corner
Stormy Bushes
Wards Plantation
Pikes Copse
Tumulus
Mill Lawn Brook
Sandy Shoot
Fire Station
Campden House
Whites Copse
Sweets Copse
Ashen Wood
Rubbles
Church Moor
Castle Hill
Fort
High Croft Wood
School
Cricket Grnd
Club House
Miracle Trees
Golf Course
Pigsty Hill
Turf
Hotel
MILL LANE
LYNDHURST ROAD
BURLEY LAWN
CHAPEL LANE
BEECHWOOD LANE
CHURCH LANE
BISTERNE CLOSE
SOUTHFIELD LA
SOUTHFIELD LANE
BENNETTS LANE
STATION RD
COTT LANE
ROAD
RINGWOOD ROAD
HONEY LANE
LONGMEAD ROAD
TYRRELLS LA
THE BARRACKS
FORRES T
COACH HILL LA
RANDALLS LANE
RINGWOOD LANE
CASTLE HILL
THE CROSS
THE MALL
DOVEYS CL
HOWARD CL
GARDEN ROAD
ESDAILE LA
CLOUGH LA
COPSE ROAD
WARNES LANE
MEADOW CL
SHAPPEN HILL LA
MOORHILL ROAD
POUND LANE
1
2
3
4
A
B
C
D
E
F

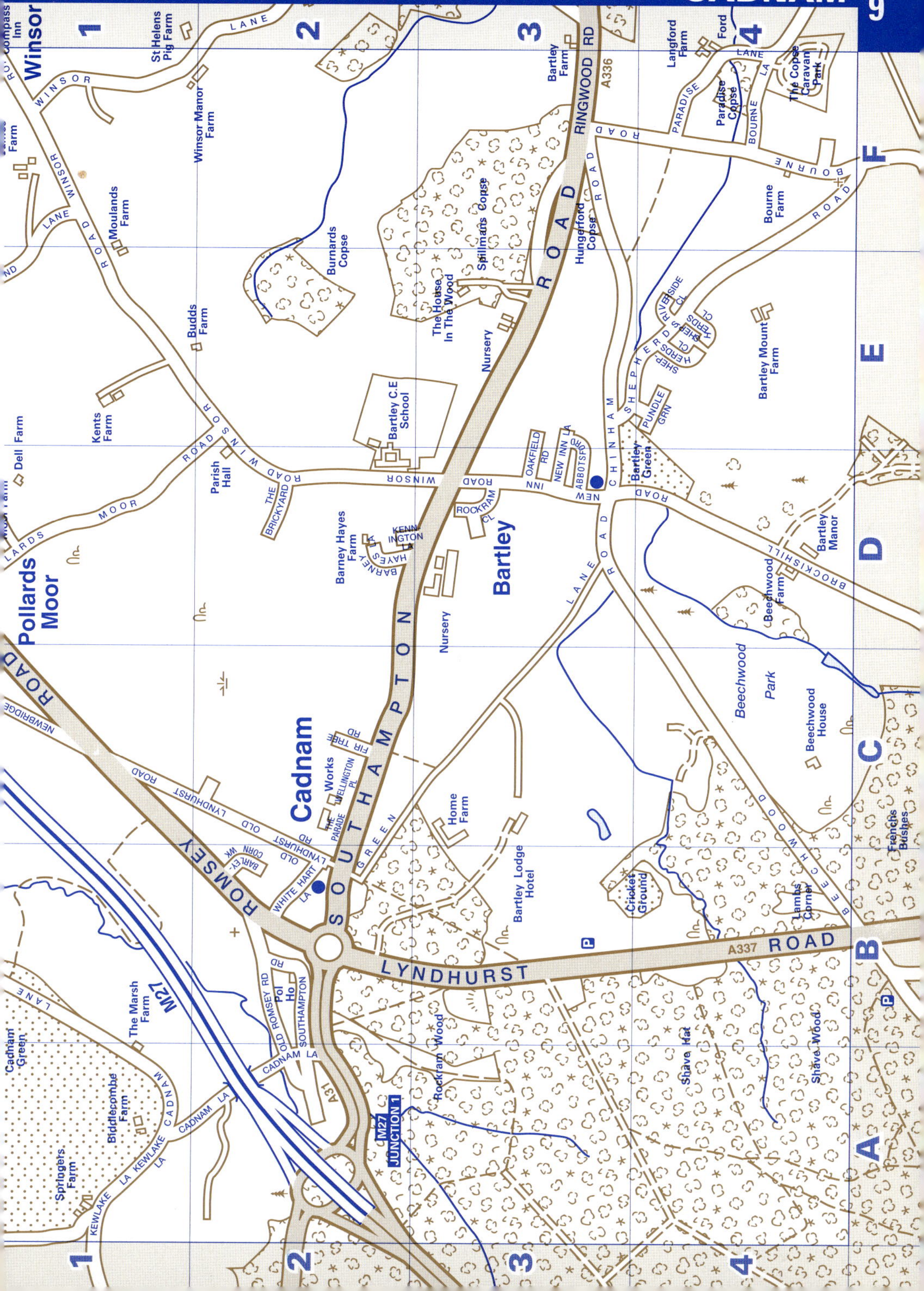
Winsor
Compass Inn
St Helens Pig Farm
Winsor Manor Farm
Moulands Farm
Budds Farm
Kents Farm
Dell Farm
Parish Hall
Pollards Moor
Burnards Copse
Spillmans Copse
The House In The Wood
Nursery
Bartley C.E School
Barney Hayes Farm
Bartley
Bartley Farm
Hungerford Copse
Langford Farm
Ford
Paradise Copse
The Copse Caravan Park
Bourne Farm
Bartley Mount Farm
Bartley Green
Bartley Manor
Beechwood Farm
Beechwood Park
Beechwood House
Frenchs Bushes
Lambs Corner
Cricket Ground
Bartley Lodge Hotel
Home Farm
Cadnam
Works
Shave Hat
Shave Wood
Rockram Wood
The Marsh Farm
Cadnam Green
Biddlecombe Farm
Springers Farm
WINSOR ROAD
WINSOR LANE
MOOR ROAD
POLLARDS MOOR ROAD
THE BRICKYARD
RINGWOOD RD
A336
ROAD
SOUTHAMPTON ROAD
ROMSEY ROAD
NEWBRIDGE ROAD
OLD LYNDHURST ROAD
LYNDHURST ROAD
A337
CHINHAM ROAD
NEW INN LA
NEW INN ROAD
OAKFIELD RD
ABBOTSFD
ROCKRAM CL
KENNINGTON LA
BARNEY HAYES LA
SHEPHERDS CL
RIVERSIDE CL
PUNDLE GRN
PARADISE LANE
BOURNE LA
BOURNE ROAD
BROCKISHILL ROAD
BEECHWOOD
LANE ROAD
GREEN
FIR TREE RD
WELLINGTON PL
THE PARADE
BARLEY CORN WK
WHITE HART LA
OLD ROMSEY RD
SOUTHAMPTON
CADNAM LA
KEWLAKE LA
CADNAM LANE
A31
M27
M27 JUNCTION 1
1 2 3 4
A B C D E F

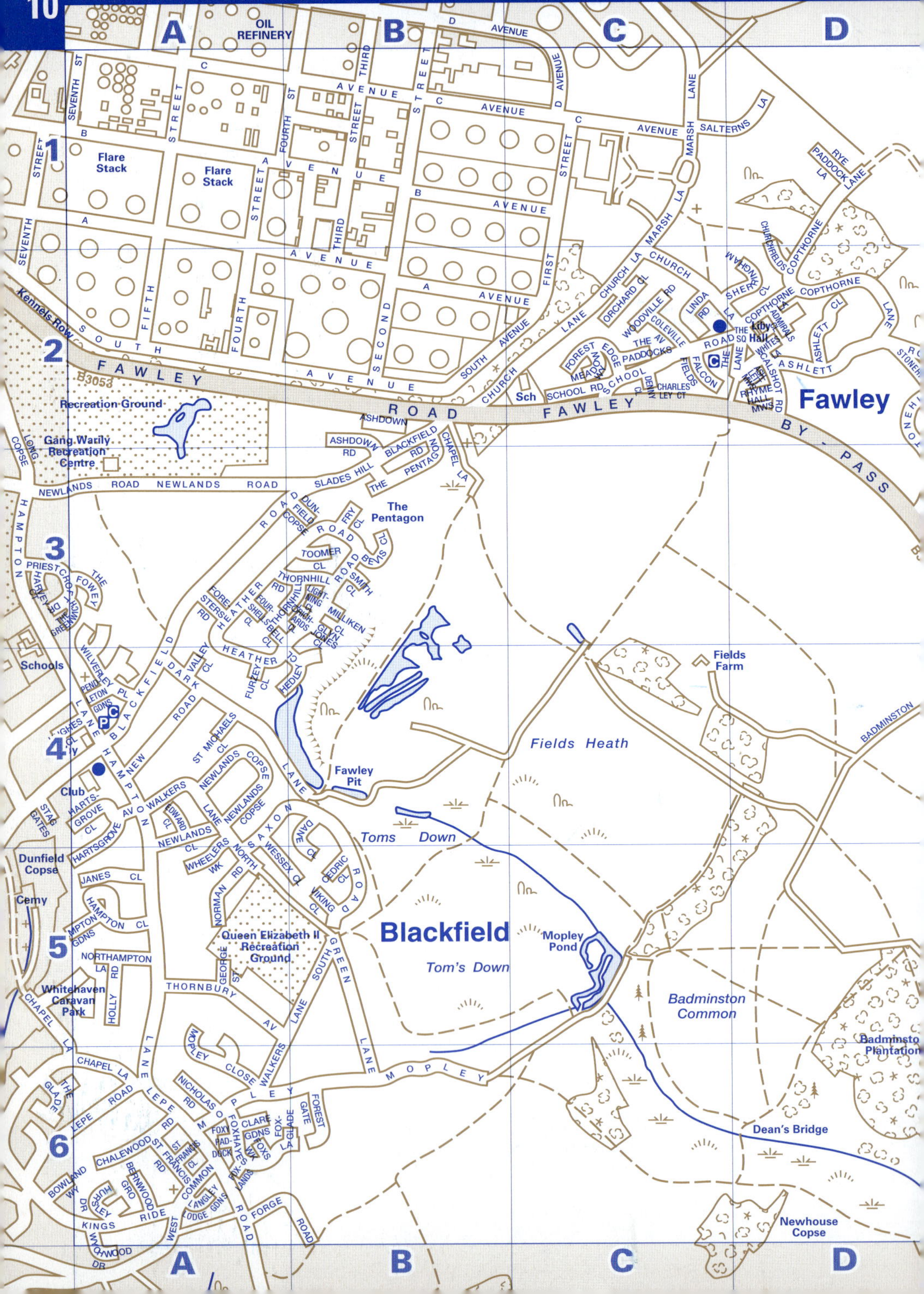
A
B
C
D
1
2
3
4
5
6
OIL REFINERY
Flare Stack
Flare Stack
SEVENTH ST
SEVENTH
STREET
FIFTH
STREET
FOURTH
FOURTH ST
THIRD
STREET
THIRD
STREET
SECOND
FIRST
STREET
AVENUE
D AVENUE
C AVENUE
B AVENUE
A AVENUE
SOUTH AVENUE
Kennels Row
SOUTH
FAWLEY
ROAD
B3053
Recreation Ground
Gang Warily Recreation Centre
LONG COPSE
NEWLANDS ROAD
HAMPTON LANE
MARSH LA
MARSH LANE
SALTERNS LA
RYE PADDOCK LANE
CHURCHFIELDS
COPTHORNE
COPTHORNE LANE
ASHLETT
ASHLETT CL
CALSHOT RD
ADMIRALS
CHURCH LA
CHURCH LANE
ORCHARD CL
WOODVILLE RD
COLEVILLE
LINDA RD
THE SQ
Lib
Hall
THE PADDOCKS
FOREST MEADOW
SCHOOL RD
SCHOOL LA
FIELDS
FALCON
CHARLES LEY CT
RHYME HALL MWS
Sch
Fawley
BY - PASS
STONEHAM
ASHDOWN
ASHDOWN RD
BLACKFIELD RD
CHAPEL LA
SLADES HILL
THE PENTAGON
The Pentagon
DUNFIELD
COPSE ROAD
FRY CL
BEVIS CL
TOOMER CL
SMITH CL
THORNHILL RD
MILLIKEN CL
HEATHER RD
FORESTERS RD
FURZEY CL
HEDLEY CL
DARK LANE
VALLEY CL
BLACKFIELD ROAD
PRIESTCROFT
HARVEY CT
THE FOWEY
Schools
WILVERLEY PL
Club
STAG GATES
HARTSGROVE CL
HARTSGROVE AV
WALKERS
ST MICHAELS CL
NEWLANDS COPSE
EDWARD CL
NEWLANDS CL
WHEELERS WK
NORTH RD
SAXON WESSEX CL
DANE CL
CEDRIC CL
VIKING CL
NORMAN RD
Fawley Pit
Toms Down
Fields Heath
Fields Farm
BADMINSTON
Blackfield
Tom's Down
Mopley Pond
Badminston Common
Badminston Plantation
Dean's Bridge
Newhouse Copse
Dunfield Copse
Cemy
JANES CL
HAMPTON GDNS
HAMPTON CL
NORTHAMPTON LA
HOLLY RD
Whitehaven Caravan Park
Queen Elizabeth II Recreation Ground
GEORGE ST
SOUTH GREEN LANE
THORNBURY AV
CHAPEL LA
THE GLADE
LEPE ROAD
LEPE RD
MOPLEY
MOPLEY CLOSE
WALKERS
NICHOLAS RD
CLARE GDNS
FOXGLADE
FOREST GATE
FOXHAYES
CHALEWOOD RD
ST FRANCIS
BERNWOOD GRO
BOWLAND WY
HURSLEY DR
KINGS RIDE
COMMON
LANGLEY LODGE GDNS
WEST
FORGE ROAD
WYCHWOOD DR

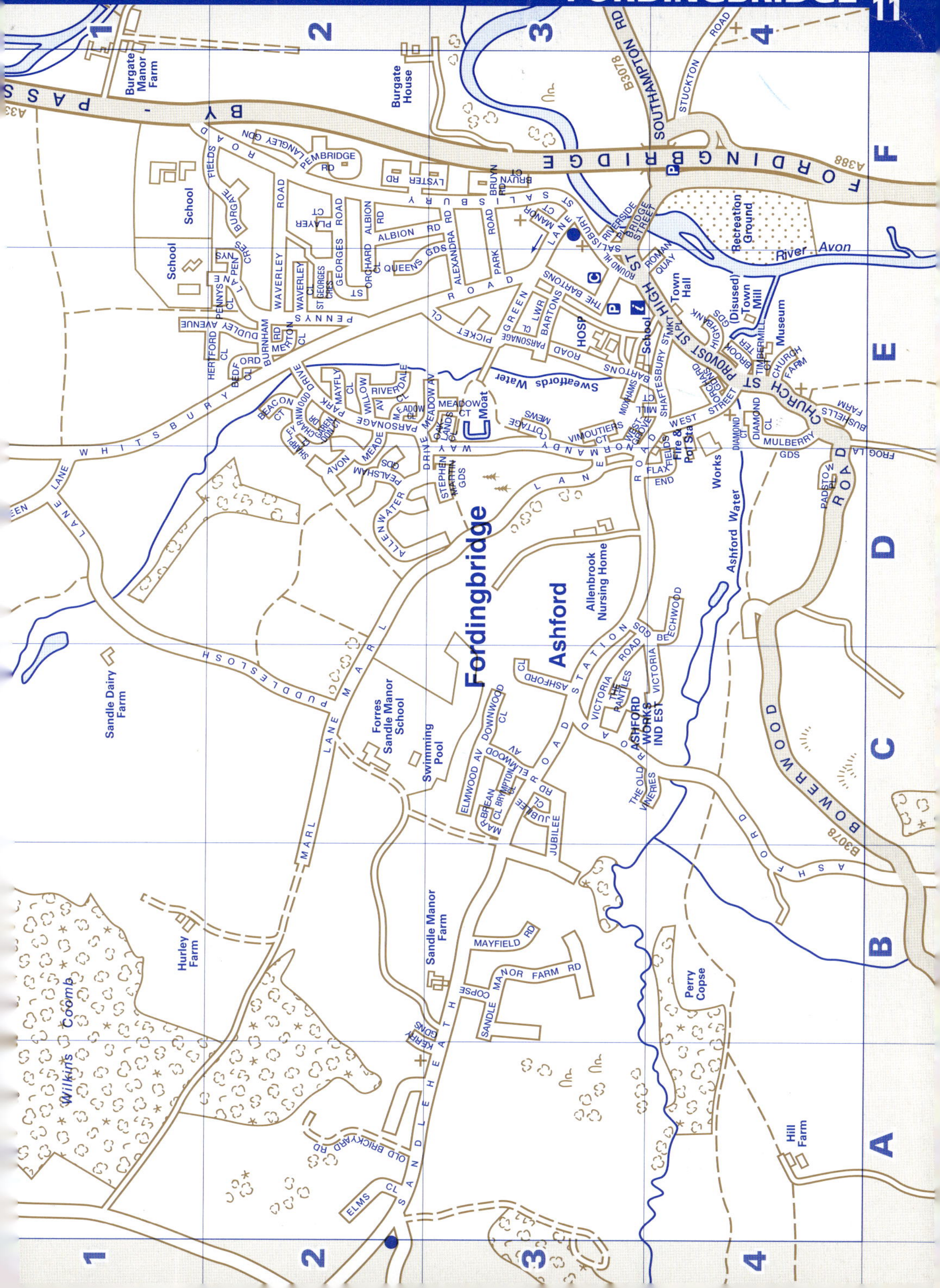
Fordingbridge
Ashford
Burgate Manor Farm
Burgate House
School
School
River Avon
Recreation Ground
Town Hall
(Disused) Town Mill
Museum
HOSP
School
Sweatfords Water
Moat
Fire & Pol Sta
Works
Ashford Water
Allenbrook Nursing Home
ASHFORD WORKS IND EST
Forres Sandle Manor School
Swimming Pool
Sandle Dairy Farm
Sandle Manor Farm
Hurley Farm
Wilkins Coomb
Perry Copse
Hill Farm
FORDINGBRIDGE BY-PASS
A338
A388
B3078
SOUTHAMPTON RD
STUCKTON ROAD
SALISBURY ROAD
HIGH ST
PROVOST ST
CHURCH ST
SHAFTESBURY ST
WEST STREET
BRIDGE STREET
ROMAN QUAY
ROUND HL
RIVERSIDE PL
ALBION RD
ALEXANDRA RD
PARK ROAD
WAVERLEY ROAD
ST GEORGES ROAD
PENNYS LANE
DUDLEY AVENUE
WHITSBURY ROAD
NORMANDY WAY
PARSONAGE PARK DRIVE
MARL LANE
PUDDLESLOSH LANE
STATION ROAD
VICTORIA ROAD
ASHFORD ROAD
ASHFORD CL
DOWNWOOD CL
ELMWOOD AV
JUBILEE RD
SANDLEHEATH ROAD
OLD BRICKYARD RD
ELMS CL
MAYFIELD RD
SANDLE MANOR FARM RD
BOWERWOOD ROAD
FROG LA
BUSHELLS FARM
MULBERRY GDS
DIAMOND CL
PADSTOW PL
ALLENWATER
AVON MEADE
PELSHAM GDS
STEPHEN MARTIN GDS
FLAX FIELDS END
VIMOUTIERS CT
COTTAGE MEWS
MEADOW CT
RIVERDALE CL
WILLOW CL
MAYFLY
BEACON CT
PICKET CL
PARSONAGE ROAD
GREEN LANE
THE BARTONS
LWR BARTONS
MANOR CT
BRUYN RD
BRUYN CT
LYSTER RD
PEMBRIDGE RD
LANGLEY GDN
FIELDS ROAD
BURGATE

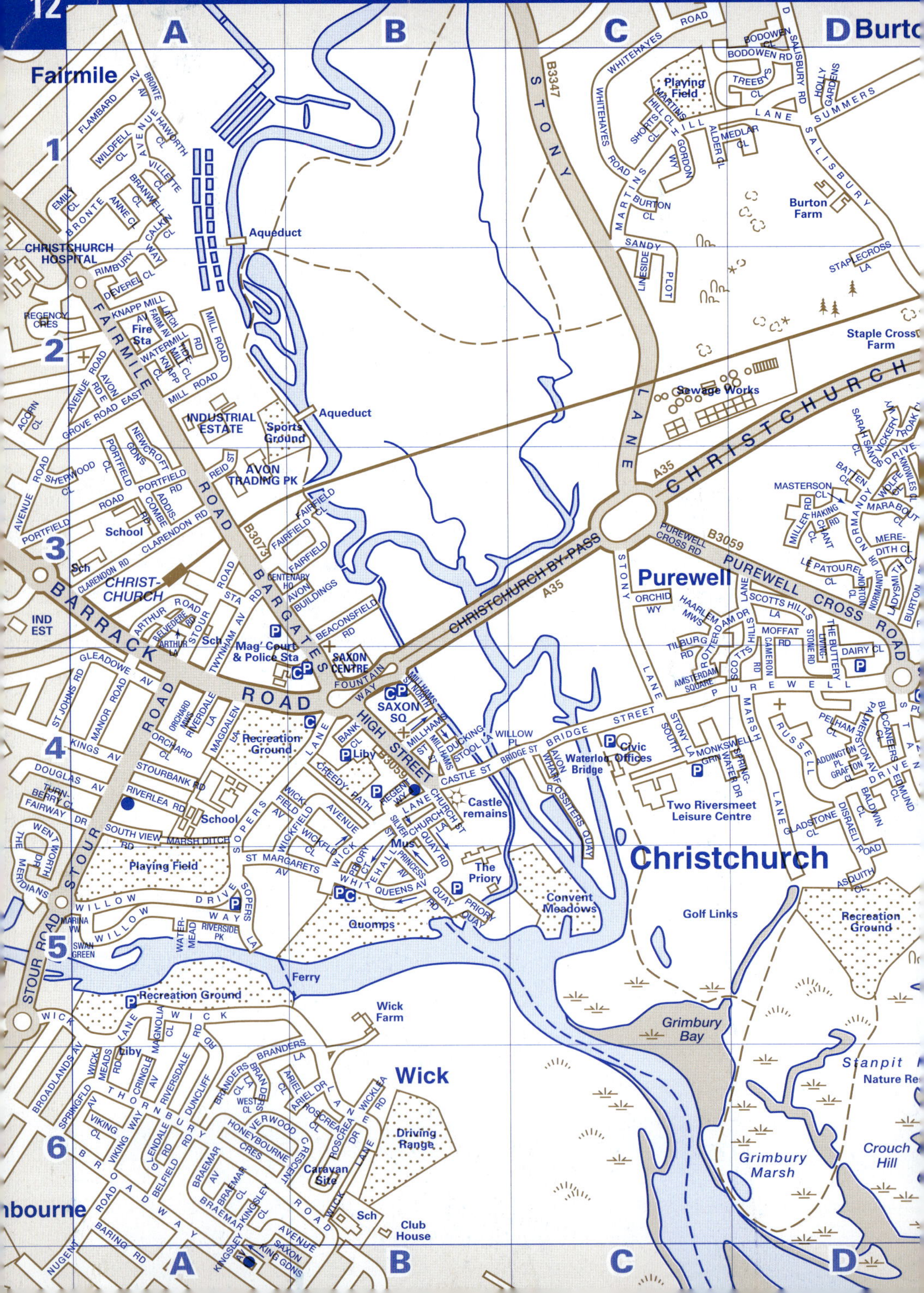
A
B
C
D
Burton
1
2
3
4
5
6
Fairmile
Flambard Av
Bronte Av
Haworth Cl
Wildfell Cl
Villette Cl
Branwell Cl
Anne Cl
Emily Cl
Bronte Av
Calkin Cl
Christchurch Hospital
Rimbury Way
Deverel Cl
Regency Cres
Knapp Mill Av
Fire Sta
Farm Av
Latch Farm Av
Watermill Rd
Tide Mill Cl
Knapp Cl
Mill Road
Fairmile Road
Acorn Cl
Avenue Road
Avon Rd East
Grove Road
Industrial Estate
Sports Ground
Aqueduct
Aqueduct
Newcroft Gdns
Portfield Cl
Portfield Rd
Addiscombe Rd
Reid St
Avon Trading Pk
Sherwood Cl
Road
School
Clarendon Rd
Portfield Av
Avenue Road
Sch
Christchurch
Bargates
B3073
Fairfield Cl
Fairfield
Fairfield
Centenary Ho
Avon Buildings
Beaconsfield Rd
Sta Rd
Barrack Road
Ind Est
Arthur Road
Belvedere Rd
Stour Rd
Arthur La
Twynham Av
Sch
Mag' Court & Police Sta
Saxon Centre
Fountain Way
Gleadowe Av
St Johns Rd
Manor Road
Orchard Mws
Riverdale La
Magdalen La
Recreation Ground
Kings Av
Orchard Cl
Stourbank Rd
Douglas Av
Turnberry Cl
Fairway Dr
Riverlea Rd
Road
Stour Road
School
South View Rd
Marsh Ditch
Soper's La
Wick Field Av
Wickfield Av
Wickfield Cl
Wick Ct
Priory Ct
Creedy Path
Avenue
Bank Cl
Liby
High Street
B3059
Saxon Sq
Millhams St
Millhams La
Ducking Stool La
Willow Pl
Castle St
Bridge St
Church St
Church La
Regent Wy
Silver St
Whitehall
Princess Av
Quay Rd
Mus
Castle remains
The Priory
Queens Av
Priory Quay
Quay Rd
St Margarets Av
Playing Field
Willow Drive
Willow Way
Water-Mead
Riverside Pk
Quomps
Marina Vw
Swan Green
The Meridians
Wentworth Dr
Ferry
Recreation Ground
Wick Lane
Wick Farm
Magnolia Cl
Wick Rd
Liby
Broadlands Av
Wick Meads Rd
Springfield Av
Cringle Av
Riversdale Rd
Duncliff Rd
Branders La
Branders Cl
Westby Cl
Ariel Cl
Ariel Dr
Roscrea Cl
Roscrea Dr
Wickfield Av
Honeybourne Cres
Verwood Crescent
Viking Cl
Viking Way
Thornbury Rd
Glendale Rd
Belfield Rd
Braemar Av
Braemar Cl
Kingsley Cl
Caravan Site
Driving Range
Wick
Sch
Club House
Broadway
Baring Rd
Nugent Rd
Kingsley Av
Saxon King Gdns
Avenue
Wick La
Stony Lane
B3347
Whitehayes Road
Whitehayes Cl
Road
Playing Field
Martins Hill Cl
Shorts Cl
Gordon Wy
Alder Cl
Medlar Cl
Martins Hill Lane
Treebys Cl
Bodowen Cl
Bodowen Rd
Salisbury Rd
Holly Gardens
Summers Lane
Salisbury Road
Burton Cl
Burton Farm
Sandy Plot
Lineside
Staplecross La
Staple Cross Farm
Sewage Works
Christchurch By-Pass
A35
A35
Christchurch
Sarah Sands Cl
Vickery Wy
Stroak
Batten Cl
Drive
Masterson Cl
Haking Rd
Miller Rd
Chant Cl
Normandy Dr
Wolfe Cl
Knowles Cl
Marabout Cl
Meredith Cl
Le Patourel Cl
Norton Cl
Normandy Dr
Ladysmith Cl
Burton
Purewell Cross Rd
B3059
Purewell Cross Road
Purewell
Stony Lane
Orchid Wy
Haarlem Mws
Rotterdam Dr
Scotts Hills La
Moffat Rd
Tilburg Rd
Cameron Rd
Scotts Hills La
Stone Rd
Livingstone Rd
The Buttery
Dairy Cl
Amsterdam Square
Purewell
Stony La South
Bridge Street
Civic Offices
Waterloo Bridge
Avon Wharf
Rossiters Quay
Monkswell Grn
Marsh La
Springwater Dr
Two Riversmeet Leisure Centre
Russell Lane
Pelham Cl
Palmerston Av
Buccaneers Cl
Addington Pl
Grafton Cl
Drive
Stanpit
Edmund Cl
Baldwin Cl
Disraeli Road
Gladstone Cl
Asquith Cl
Christchurch
Golf Links
Recreation Ground
Convent Meadows
Grimbury Bay
Stanpit
Nature Re
Grimbury Marsh
Crouch Hill
bourne

E
F
G
H
1
2
3
4
5
6
BY - PASS
LYNDHURST ROAD
ROESHOT HILL
HIGHCLIFFE ROAD
SOMERFORD ROAD
MUDEFORD
Somerford
Stanpit
Friars Cliff
Mudeford
Staple Cross
Sch
School
Playing Field
Recreation Ground
Super-store
Somerford Bridge
Sports Ground
Caravan Site
Hoburne Caravan Park
SEA VIXEN IND EST
GRANGE RD BSNS CENTRE
PRIORY INDUSTRIAL PARK
SOMERFORD BSNS PK
HUGHES BSNS CENTRE
INDUSTRIAL ESTATE
SILVER BSNS PK
BEAVER IND EST
AIRFIELD IND EST
AMBASSADOR IND EST
Factory
Comm Centre
Rec Grnd
Caravan Park
Sandhills
River Mude
Mean High Water
Mean Low Water
Little Haven
LB Station
The Run
Blackberry Point
Groynes
14
WATERY LANE
STANPIT
FISHERMANS BANK
PROMENADE
SOUTHCLIFFE ROAD
BURE LANE
BURE HOMAGE LANE
HOBURNE LANE
SAFFRON DRIVE
HUNT ROAD
DORSET ROAD
DRUITT ROAD
AMETHYST ROAD
EDWARD ROAD
RAVEN WAY
HAVEN DRIVE
DEHAVILLAND WAY
RUNWAY

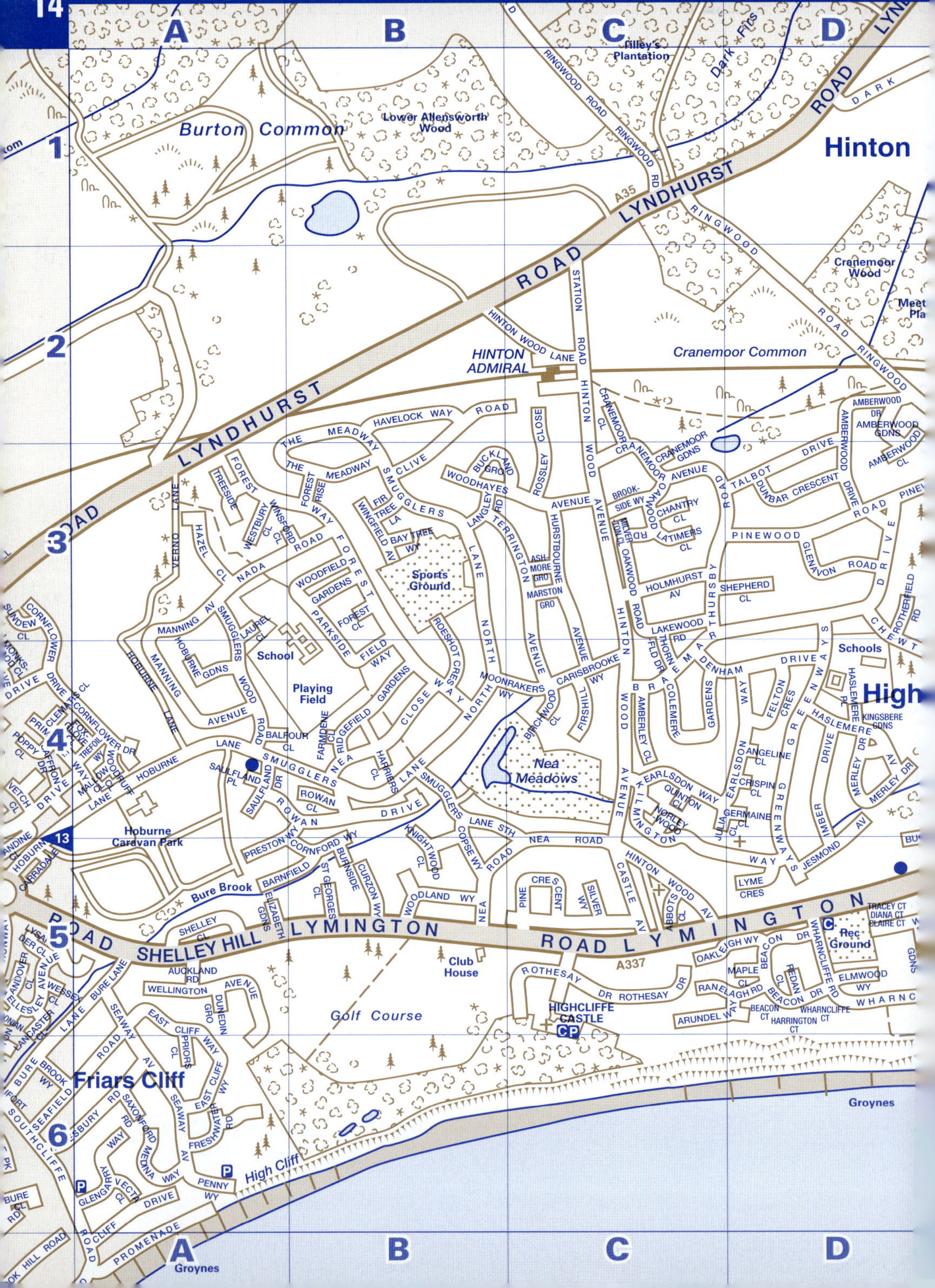
A
B
C
D
1
2
3
4
5
6
Tilley's Plantation
Dark Firs
Burton Common
Lower Allensworth Wood
Hinton
Cranemoor Wood
Meet Pla
Cranemoor Common
HINTON ADMIRAL
LYNDHURST ROAD
A35
RINGWOOD ROAD
RINGWOOD RD
DARK
STATION ROAD
HINTON WOOD LANE
HINTON WOOD AVENUE
ROSSLEY CLOSE
CRANEMOOR CL
CRANEMOOR AVENUE
CRANEMOOR GDNS
AMBERWOOD DR
AMBERWOOD GDNS
AMBERWOOD CL
AMBERWOOD DRIVE
TALBOT DRIVE
DUNBAR CRESCENT
PINEWOOD ROAD
GLENAVON ROAD
ROTHERFIELD RD
CHEWTON
HAVELOCK WAY
THE MEADWAY
CLIVE ROAD
SMUGGLERS LANE
FIR TREE LA
BAY TREE WY
WINGFIELD AV
WOODHAYES AVENUE
BUCKLAND GRO
LANGLEY RD
TERRINGTON AVENUE
HURSTBOURNE AVENUE
ASHMORE GRO
MARSTON GRO
OAKWOOD ROAD
CHANTRY CL
LATIMERS CL
BROOK-SIDE WY
HOLMHURST AV
SHEPHERD CL
THURSBY ROAD
LAKEWOOD RD
DENHAM DRIVE
Sports Ground
FOREST WAY
FOREST RISE
TREESIDE
WESTBURY CL
WINSFORD CL
HAZEL CL
NADA ROAD
VERNO LANE
WOODFIELD GARDENS
FOREST CL
PARKSIDE
FIELD WAY
ROESHOT CRES
NORTH AVENUE
MOONRAKERS WY
CARISBROOKE WY
BIRCHWOOD CL
FIRSHILL
Nea Meadows
School
Playing Field
Schools
High
KINGSBERE GDNS
HASLEMERE AV
HASLEMERE PL
MERLEY DR
MERLEY AV
FELTON CRES
GREENWAYS
ANGELINE CL
CRISPIN CL
GERMAINE CL
JULIA CL
EARLSDON WAY
QUINTON CL
NORLEY WOOD
KILMINGTON WAY
JESMOND AV
IMBER DRIVE
LYME CRES
AMBERLEY CL
COLEMERE GARDENS
THORNEMAN CL
MANNING AV
HOBURNE GDNS
SMUGGLERS WOOD ROAD
LAUREL CL
HOBURNE LANE
MANNING LANE
HOBURNE
BALFOUR CL
FARMDENE CL
RIDGEFIELD GARDENS
CLOSE
HARRIERS CL
SAULFLAND PL
SAULFLAND DR
ROWAN CL
ROWAN DRIVE
PRESTON WY
CORNFORD WY
BURNSIDE
CURZON WY
KNIGHTWOOD CL
COPSE WY
NEA ROAD
SMUGGLERS LANE STH
Hoburne Caravan Park
13
CORNFLOWER DR
SUNDEW CL
MONKS CL
CLEMATIS CL
PRIMROSE
FOXGLOVE
TREFOIL WY
MALLOW CL
WOODRUFF
POPPY CL
SAFFRON DR
VETCH CL
HOBURN
CARRADALE
Bure Brook
BARNFIELD
ST GEORGES CL
ELIZABETH GDNS
WOODLAND WY
PINE CRESCENT
SILVER WY
CASTLE AV
ABBOTS CL
HINTON WOOD AV
ROAD
SHELLEY HILL
SHELLEY CL
LYMINGTON ROAD
A337
TRACEY CT
DIANA CT
CLAIRE CT
Rec Ground
OAKLEIGH WY
BEACON DR
WHARNCLIFFE RD
MAPLE CL
REDAN CL
ELMWOOD WY
RANELAGH RD
BEACON CT
HARRINGTON CT
WHARNCLIFFE CT
ARUNDEL WAY
Club House
ROTHESAY DR
HIGHCLIFFE CASTLE
CP
Golf Course
AUCKLAND AVENUE
WELLINGTON AVENUE
DUNEDIN GRO
EAST CLIFF WAY
PRIORS CL
EAST CLIFF WAY
BURE LANE
SEAWAY AV
ANDOVER
LYSANDER CL
WESSEX AVENUE
KELLESLEY CL
LANCASTER CL
BURE BROOK WY
SEAFIELD ROAD
Friars Cliff
SAXONFORD RD
SEAWAY
FRESHWATER DR
MEDINA WAY
GLENGARRY WAY
VECTA CL
DRIVE
PENNY WY
P
High Cliff
SOUTHCLIFFE
CLIFF
BURE RD
PROMENADE
Groynes

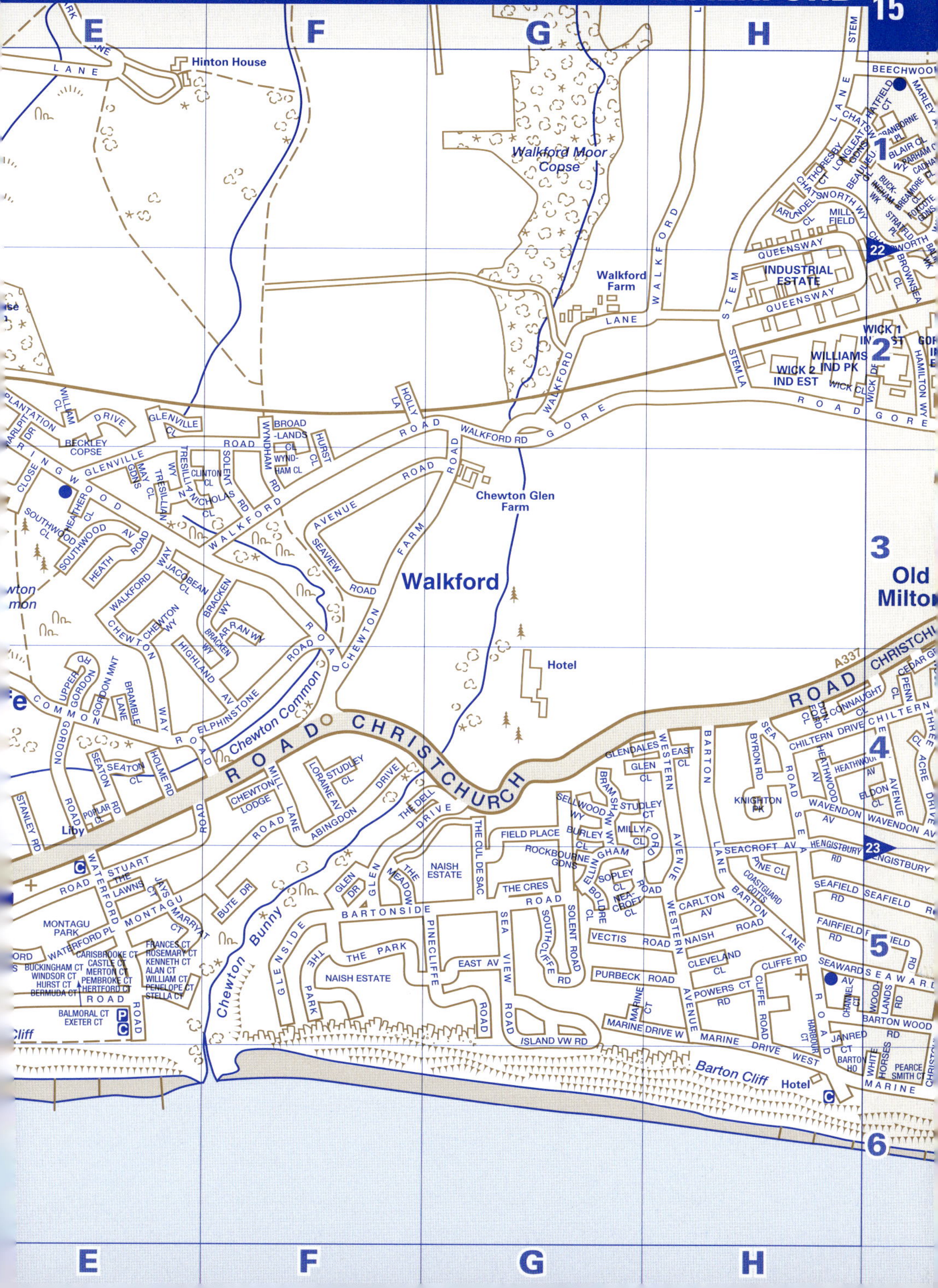
E
F
G
H
1
2
3
4
5
6
Hinton House
LANE
Walkford Moor Copse
Walkford Farm
WALKFORD LANE
STEM LANE
STEM LA
QUEENSWAY
INDUSTRIAL ESTATE
WICK 1 IND EST
WICK 2 IND EST
WILLIAMS IND PK
WICK CL
WICK DR
HAMILTON WY
GORE ROAD
BEECHWOOD
MARLEY AV
HATFIELD CT
CHATSWORTH WY
LONGLEAT GDNS
BEAULIEU CL
THORESBY CT
ARUNDEL CL
MILL-FIELD
BLAIR CL
PARHAM CL
CRANBORNE PL
BUCKINGHAM WK
BREAMORE CL
FOXCOTE GDNS
STRATFIELD PL
BROWNSEA CL
22
GLENVILLE CL
GLENVILLE
BECKLEY COPSE
PLANTATION DR
WILLIAM CL
DRIVE
RINGWOOD ROAD
CLOSE
BROADLANDS CL
WYNDHAM RD
WYNDHAM CL
HURST CL
SOLENT RD
CLINTON CL
NICHOLAS CL
TRESILLIAN WY
TRESILLIAN CL
MAY GDNS
HOLLY LA
WALKFORD RD
WALKFORD ROAD
FARM ROAD
AVENUE ROAD
SEAVIEW ROAD
Chewton Glen Farm
Walkford
Hotel
SOUTHWOOD CL
SOUTHWOOD AV
HEATHER CL
HEATH ROAD
WALKFORD WAY
JACOBEAN CL
BRACKEN WY
ARRAN WY
CHEWTON WY
CHEWTON COMMON ROAD
HIGHLAND AV
ELPHINSTONE ROAD
Chewton Common
UPPER GORDON RD
GORDON MNT
BRAMBLE LANE
GORDON ROAD
SEATON CL
SEATON RD
HOLME RD
POPLAR CL
POPLAR RD
STANLEY RD
Liby
A337
CHRISTCHURCH ROAD
MILL LANE
CHEWTON LODGE
LORAINE AV
STUDLEY CL
ABINGDON DRIVE
ABINGDON ROAD
THE DELL
Chewton Bunny
STUART ROAD
THE LAWNS
JAYS CT
MARRYAT CT
MONTAGU ROAD
MONTAGU PARK
WATERFORD ROAD
WATERFORD PL
CARISBROOKE CT
CASTLE CT
MERTON CT
HERTFORD CT
BUCKINGHAM CT
WINDSOR CT
HURST CT
BERMUDA CT
FRANCES CT
ROSEMARY CT
KENNETH CT
ALAN CT
WILLIAM CT
PENELOPE CT
STELLA CT
BALMORAL CT
EXETER CT
BUTE DR
GLEN DR
GLENSIDE
THE PARK
THE MEADOW
NAISH ESTATE
BARTONSIDE ROAD
PINECLIFFE
EAST AV
THE CUL DE SAC
FIELD PLACE
ROCKBOURNE GDNS
THE CRES
SEA VIEW ROAD
SOUTHCLIFFE RD
SOLENT ROAD
ISLAND VW RD
GLENDALES
GLEN CL
EAST CL
WESTERN AVENUE
SELLWOOD WY
BRAMSHAW WY
STUDLEY CT
BURLEY CL
MILLYFORD CL
ELLINGHAM
BOLDRE CL
SOPLEY CL
NEACROFT CL
VECTIS ROAD
NAISH ROAD
PURBECK ROAD
CLEVELAND CL
MARINE CT
MARINE DRIVE W
POWERS CT
POWERS RD
CLIFFE ROAD
CLIFFE RD
CARLTON AV
BARTON LANE
BARTON COURT ROAD
COASTGUARD COTTS
MARINE DRIVE WEST
Barton Cliff
Hotel
BYRON RD
KNIGHTON PK
SEA ROAD
SEACROFT AV
PINE CL
DUNFORD CL
CONNAUGHT CL
CHILTERN DRIVE
HEATHWOOD AV
HEATHWOOD
ELDON CL
WAVENDON AV
AVENUE DRIVE
ACRE
HENGISTBURY RD
23
SEAFIELD RD
FAIRFIELD RD
SEAWARD AV
HARBOUR CT
HARBOUR ROAD
CHANNEL CT
WOODLANDS RD
BARTON WOOD RD
JANRED CT
BARTON HO
WHITE HORSES
PEARCE SMITH CT
MARINE
CEDAR
PENN CL
THREE
Old Milton
Cliff

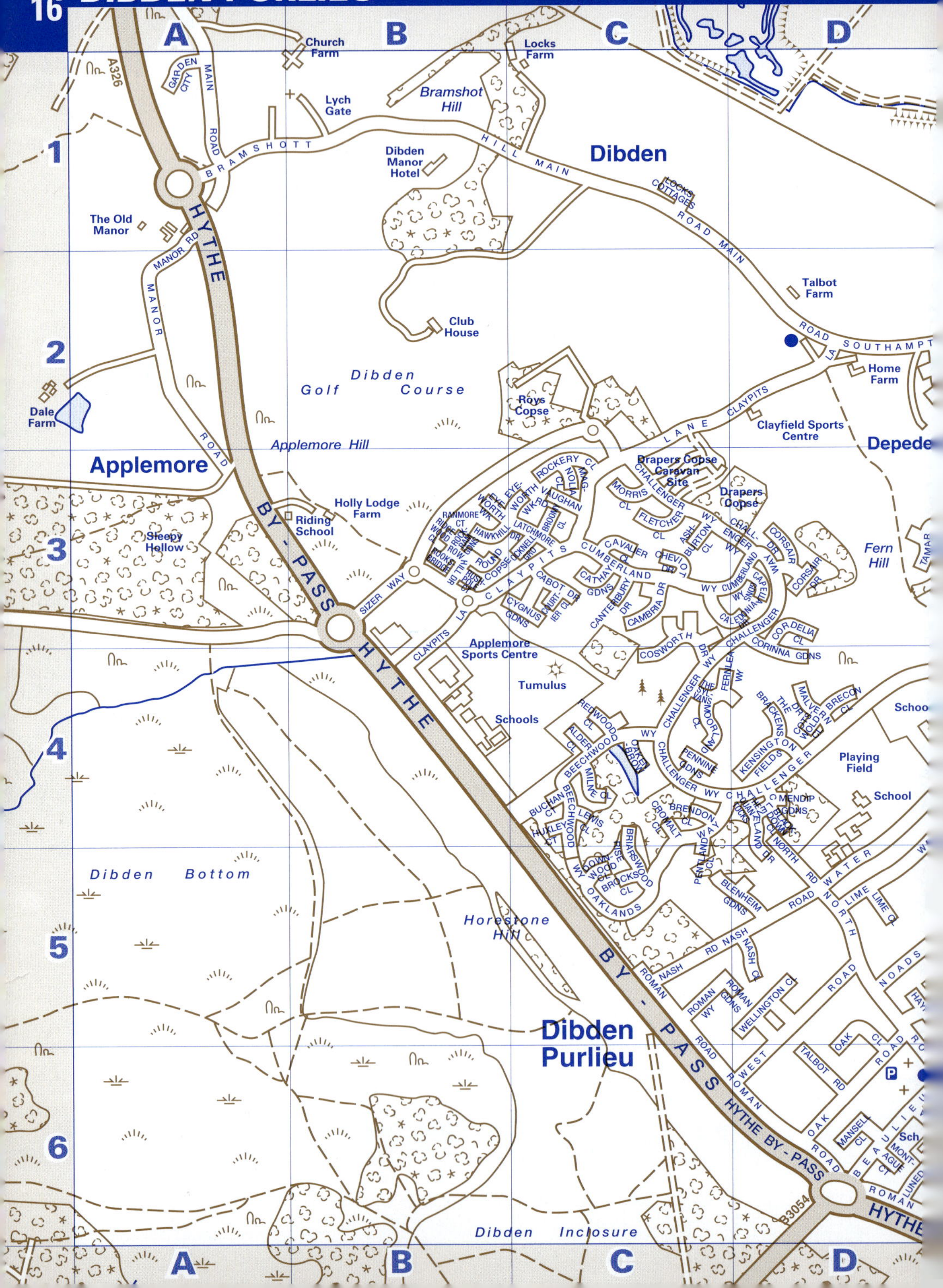

A
B
C
D
1
2
3
4
5
6
A326
GARDEN CITY
MAIN ROAD
Church Farm
Lych Gate
Locks Farm
Bramshot Hill
BRAMSHOTT
HILL
MAIN
Dibden
Dibden Manor Hotel
LOCKS COTTAGES
ROAD
MAIN
The Old Manor
MANOR RD
MANOR
HYTHE
Talbot Farm
Club House
ROAD
SOUTHAMPT
LA
Home Farm
Dibden Golf Course
Dale Farm
Roys Copse
CLAYPITS
LANE
Clayfield Sports Centre
Applemore Hill
Depede
ROAD
Applemore
Drapers Copse Caravan Site
Drapers Copse
Holly Lodge Farm
Riding School
Sleepy Hollow
BY-PASS
ROCKERY CL
MORRIS CL
CHALLENGER
FLETCHER CL
VAUGHAN
CAVALIER CL
CHEVIOT DR
CUMBERLAND
CORSAIR DR
Fern Hill
TAMAR
SIZER WAY
CLAYPITS
CABOT DR
CYGNUS GDNS
CANTERBURY DR
CAMBRIA DR
CALEDONIA
CORDELIA CL
CORINNA GDNS
COSWORTH
CLAYPITS LA
Applemore Sports Centre
Tumulus
Schools
HYTHE
REDWOOD CL
ALDER CL
BEECHWOOD
MILNE CL
CHALLENGER WY
PENNINE GDNS
KENSINGTON FIELDS
MALVERN
BRECON CL
Schoo
Playing Field
School
MENDIP GDNS
BUCHANAN CT
HUXLEY CT
LEWIS CL
BRIARSWOOD
BROCKSOOD CL
OAKLANDS
BRENDON
CROMALT CL
PENTLAND CL
BLENHEIM GDNS
NORTH
WATER
ROAD
LIME CL
Dibden Bottom
Horestone Hill
BY-PASS
ROMAN
NASH
RD NASH
NASH CL
ROMAN WY
ROMAN GDNS
WELLINGTON CL
ROAD
NOADS
Dibden Purlieu
ROAD
ROMAN
WEST
TALBOT RD
OAK
OAK
ROAD
P
MANSELL CL
BEAULIEU
MONTAGUE CT
HYTHE BY-PASS
ROMAN
HYTHE
B3054
Dibden Inclosure

E
F
G
H
1
2
3
4
5
6
Lock
White Heather Ct
Velsheda Ct
Astra Ct
Hythe Marina
Endeavour Way
Shamrock Wy
Slipway
Ferry Landing Stage
River Test
Hythe Hard
Railway
Hythe Pier
Pier
West Cliff Hall
Jessop Wk
Waterside
Mount House Cl
Pol Sta
Recreation Ground
Prospect Pl
High St
The Promenade
The Marsh
Comm Centre
School
Library
St Johns St
Sir Christopher Cl
The Shipyard Ind Est
Slipway
Pontoons
Pier
Hythe
Fire Sta
Shore Road
Dominy Cl
Spring Rd
Hythe Sailing Club
Hall
Nature Reserve
Beaumaris Gdns
West Road
Jones Lane
Dibden Lodge Cl
Southampton Road
Malwood Road
Michaels Wy
Douglas Wy
Roberts Rd
Solent Dr
Hollybank
Fairway Road
Atheling Rd
Mousehole La
Coat Gdns
Hobart Dr
Langdown
Ashford Cres
Grays Av
Seaward Rd
Sch
Harvey Gdns
Shore Road Playing Field
Tates Copse
Sewage Works
Pipers Copse
Copsewood Rd
Hillview
Fern Rd
Mead
Glenside
The Vale
Sunningdale
Whittington Close
Fairview Dr
Fairview Cl
Langdown Firs
Rec Grnd
Windrush Wy
Elgin Cl
Queens Cl
Knightwood Av
Home Farm Cl
Tates Rd
Deerleap Way
Adams Rd
Burnett Cl
Fulmar Dr
Guillemot Cl
Gannet Cl
Cormorant Dr
Curlew Dr
Langdown Lawn
Hythe & Dibden Hospital
Golden Hind Pk
Beaulieu Road
Highlands Wy
Ladycross Rd
Furzey Holts
Furzey Av
Furzey Piece
Curlew Wk
Frost Lane
Hart Hill
Frostlane Copse
Pinewood Cres
Spinney Dale
Buttsash Wood
Rosebery Av
Lammas Rd
Furzedale Gdns
Pinewood Dr
Hotel
Furzedown Farm
Frostlane
Whitewater Rise
Duke's Wood Dr
Foxtail Dr
Bullrush Cl
Butts Bridge
Elizabeth Gdns
Kitchers Copse
Seadown Veterinary Hospital
Forest Lodge
Forest Lodge Farm
Crampool Copse
Gringos Copse
Works
Charleston Rd
Avenue Rd
Fawley Rd
Chaloner Cres
Yelverton Av
Fleuret Cl
Ashleigh Cl
Buttsash
Butts Ash Lane
Crescent Road
Beech Cl
Forest Front
Forest Mdw
Blenheim Mws
Beverley Rd
Arnwood Av
Monks Wk
Work
Schs

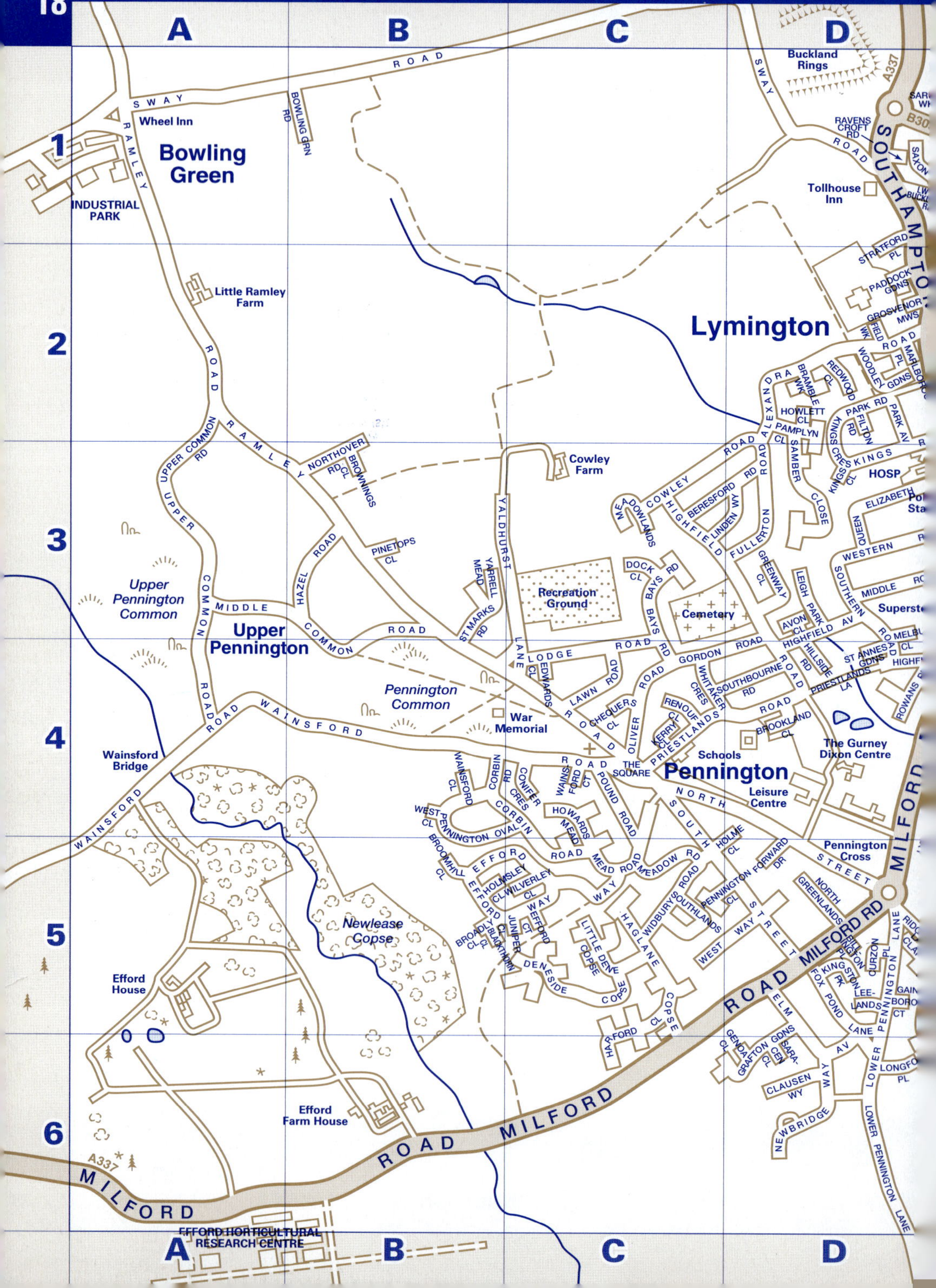
A
B
C
D
1
2
3
4
5
6
Bowling Green
Wheel Inn
INDUSTRIAL PARK
SWAY ROAD
BOWLING GRN RD
RAMLEY ROAD
Little Ramley Farm
Buckland Rings
Tollhouse Inn
RAVENS CROFT RD
SOUTHAMPTON ROAD
A337
B3058
STRATFORD PL
PADDOCK GDNS
GROSVENOR MWS
FIELD WK
WOODLEY GDNS
MARLBOROUGH
Lymington
ALEXANDRA ROAD
BRAMBLE WK
REDWOOD CL
HOWLETT CL
PAMPLYN CL
PARK RD
FILTON RD
PARK AV
KINGS CRES
KINGS CL
KINGS CLOSE
SAMBER CLOSE
HOSP
ELIZABETH
QUEEN
WESTERN
SOUTHERN
MIDDLE
Superstore
COWLEY ROAD
Cowley Farm
NORTHOVER RD
BROWNINGS CL
UPPER COMMON RD
UPPER COMMON ROAD
MEADOWLANDS
BERESFORD RD
LINDEN WY
HIGHFIELD ROAD
FULLERTON
GREENWAY CL
LEIGH PARK
AVON CL
HIGHFIELD AV
HILLSIDE RD
ST ANNES GDNS
MELBURY
PRIESTLANDS LA
ROWANS
Upper Pennington Common
Upper Pennington
HAZEL ROAD
MIDDLE COMMON ROAD
PINETOPS CL
YARRELL MEAD
YALDHURST LANE
ST MARKS RD
Recreation Ground
DOCK CL
BAYS RD
Cemetery
LODGE ROAD
LODGE CL
EDWARDS CL
LAWN ROAD
GORDON ROAD
WHITAKER CRES
SOUTHBOURNE RD
RENOUF CL
KERRY CL
PRIESTLANDS
BROOKLAND CL
The Gurney Dixon Centre
Pennington Common
War Memorial
WAINSFORD ROAD
Wainsford Bridge
CHEQUERS CL
OLIVER RD
THE SQUARE
Schools
Pennington
Leisure Centre
WAINSFORD CL
CORBIN RD
CONIFER CRES
CORBIN
WAINSFORD CT
POUND ROAD
HOWARDS MEAD
NORTH
SOUTH
HOLME CL
WEST CL
PENNINGTON OVAL
BROOMHILL
EFFORD
HOLMSLEY CL
WILVERLEY CL
WAY
MEAD ROAD
MEADOW WAY
FORWARD DR
Pennington Cross
STREET
GREENLANDS
NORTH
EFFORD CL
BROADLY CL
JUNIPER CL
BLACKTHORN
EFFORD CT
LITTLE DENE
COPSE
HAGLANE
WIDBURY RD
SOUTHLANDS
PENNINGTON CL
DENESIDE
WEST WAY
MILFORD RD
ELM
BILLINGTON PL
KINGSTON PK
CURZON PL
FOX POND LANE
LEE-LANDS
PENNINGTON
LANE
GAINSBOROUGH CT
RIDGE CL
HARFORD CL
COPSE
Newlease Copse
Efford House
GENOA CL
GRAFTON GDNS
SARACEN CL
AV
CLAUSEN WY
WAY
LOWER
LONGFORD PL
NEWBRIDGE WAY
LOWER PENNINGTON LANE
Efford Farm House
ROAD MILFORD
MILFORD ROAD
MILFORD ROAD
A337
EFFORD HORTICULTURAL RESEARCH CENTRE

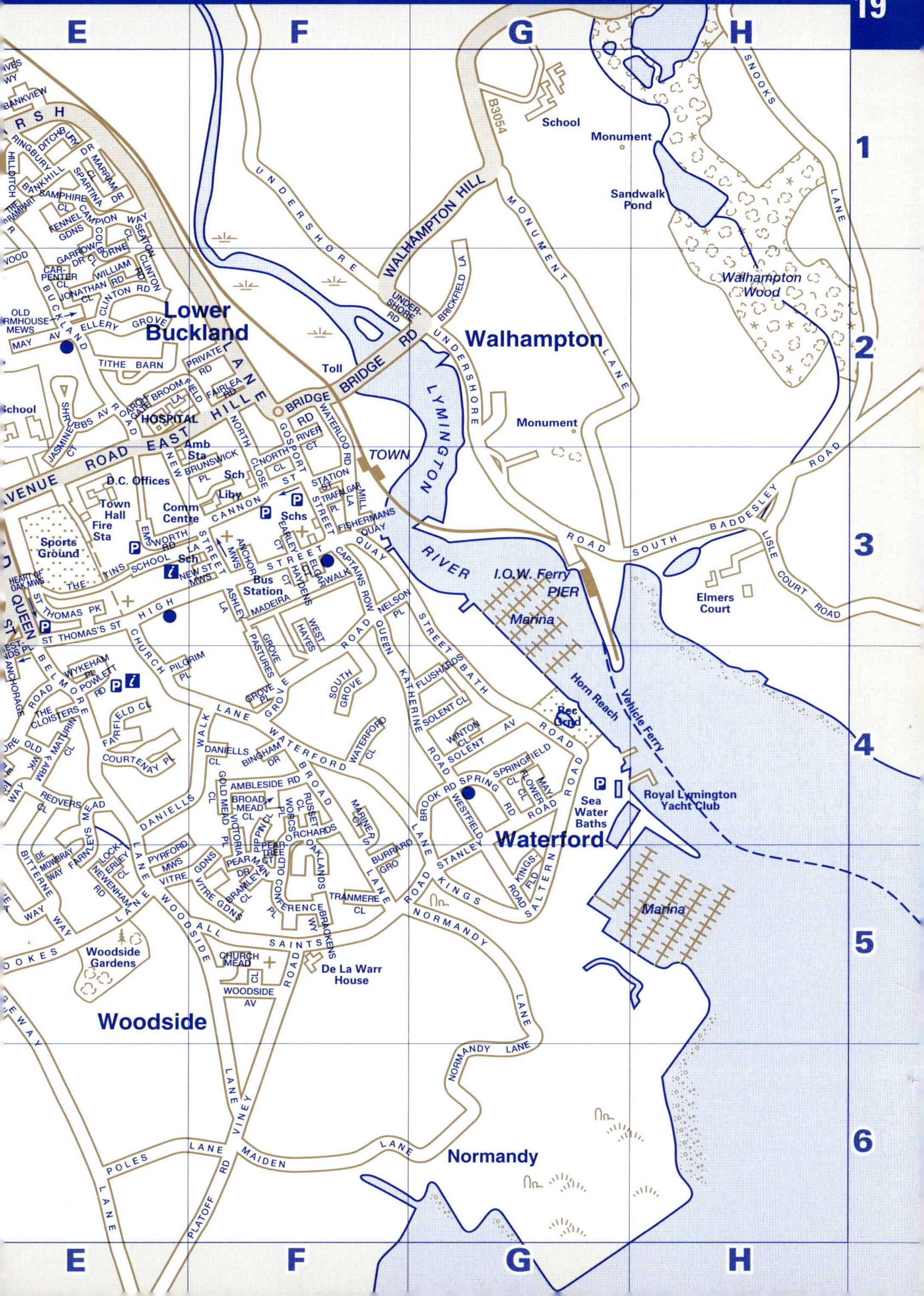
E
F
G
H
1
2
3
4
5
6
Lower Buckland
Walhampton
Waterford
Woodside
Normandy
School
Monument
Sandwalk Pond
Walhampton Wood
Monument
Toll
TOWN
LYMINGTON
RIVER
I.O.W. Ferry
PIER
Marina
Horn Reach
Vehicle Ferry
Rec Grnd
Royal Lymington Yacht Club
Sea Water Baths
Marina
Elmers Court
HOSPITAL
Amb Sta
D.C. Offices
Town Hall
Fire Sta
Sports Ground
Comm Centre
Sch
Liby
Schs
Bus Station
Woodside Gardens
De La Warr House
B3054
WALHAMPTON HILL
UNDERSHORE
UNDERSHORE RD
BRICKFIELD LA
MONUMENT LANE
BRIDGE RD
BRIDGE RD
SNOOKS LANE
BADDESLEY ROAD
SOUTH BADDESLEY ROAD
LISLE COURT ROAD
EAST HILL
AVENUE ROAD
HIGH STREET
ST THOMAS'S ST
ST THOMAS PK
QUEEN ST
BELMORE RD
CHURCH LANE
WATERFORD LANE
GROVE ROAD
QUEEN KATHERINE ROAD
BATH ROAD
KINGS SALTERN ROAD
NORMANDY LANE
MAIDEN LANE
VINEY RD
PLATOFF RD
POLES LANE
ALL SAINTS ROAD
WOODSIDE LANE
STANLEY RD
BROOK RD
SOLENT AV
FLUSHARDS
WINTON CL
SOLENT CL
SPRINGFIELD CL
FLOWER MAY
WESTFIELD RD
NELSON PL
CAPTAINS ROW
QUAY
FISHERMANS QUAY
MILL LA
STATION ST
GOSPORT ST
WATERLOO RD
RIVER CT
NORTH CL
CANNON ST
NEW ST
NEW STREET MWS
SCHOOL LA
ELGAR WALK
WEST HAYES
HAYDENS CT
MADEIRA
ASHLEY LA
GROVE PASTURES
SOUTH GROVE
GROVE PL
PILGRIM PL
DANIELLS CL
DANIELLS WALK
BINGHAM DR
AMBLESIDE RD
BROAD MEAD CL
GOLD MEAD CL
VICTORIA PL
RUSSET CL
ORCHARDS
OAKLANDS
MARINER CT
BURRARD GRO
TRANMERE CL
BRACKENS WY
CONFERENCE PL
PEAR TREE CT
BRAMLEY CL
PEARMAIN DR
PYRFORD MWS
VITRE GDNS
WOODSIDE AV
CHURCH MEAD
NEWENHAM RD
LOCKERLEY CL
FARNLEYS MEAD
REDVERS CL
MATURIN CL
FAIRFIELD CL
COURTENAY PL
WYKEHAM PL
POWLETT RD
THE CLOISTERS
ANCHORAGE
OLD FARM WK
BITTERNE WAY
DE MOWBRAY WAY
STOKES
TITHE BARN
PRIVATE RD
FAIRLEA RD
BROOM FIELD
ORCHARD GATE
SHIRLEY
GIBBS AV
JASMINE CT
BRUNSWICK PL
ELLERY GROVE
BUCKLAND
OLD FARMHOUSE MEWS
MAY AV
CARPENTER CL
JONATHAN RD
WILLIAM RD
CLINTON RD
GARROW DR
COLBORNE
SEATON
CAMPION WAY
FENNEL GDNS
SAMPHIRE CL
SPARTINA DR
MARRAM DR
BANKHILL DR
RINGBURY
DITCHBURY
HILLDITCH
THE RAMPART
MARSH
BANKVIEW
HEART OF OAK MWS
TINS
EMSWORTH RD
ANCHOR MWS
EARLEY CT
TRAFALGAR ST

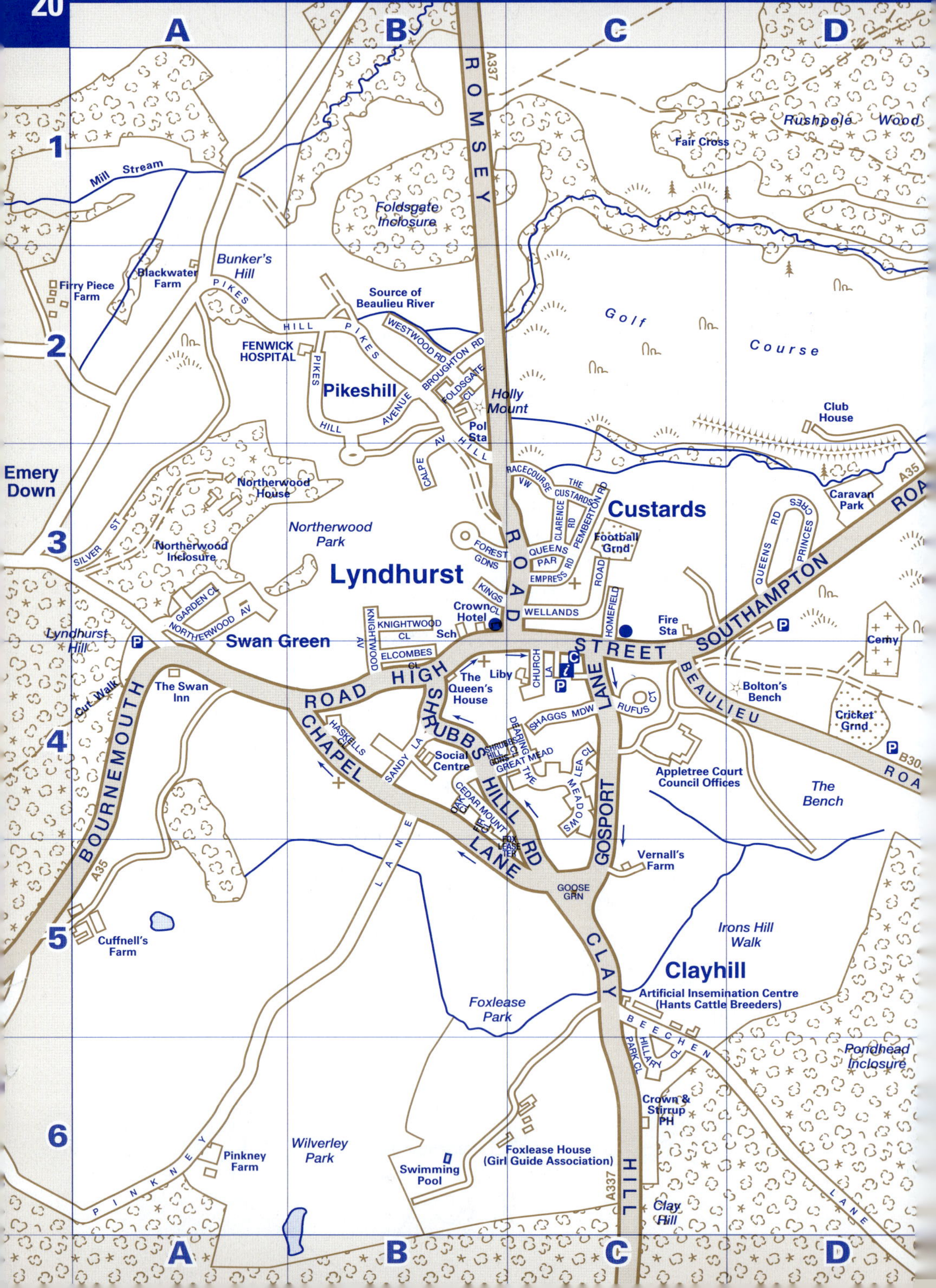
A
B
C
D
1
2
3
4
5
6
Mill Stream
Foldsgate Inclosure
Rushpole Wood
Fair Cross
Bunker's Hill
Blackwater Farm
Firry Piece Farm
Source of Beaulieu River
PIKES HILL
FENWICK HOSPITAL
Pikeshill
WESTWOOD RD
BROUGHTON RD
FOLDSGATE CL
Holly Mount
Pol Sta
AVENUE
CALPE AV
ROMSEY ROAD
A337
Golf Course
Club House
Emery Down
Northerwood House
Northerwood Park
Northerwood Inclosure
SILVER ST
GARDEN CL
NORTHERWOOD AV
RACECOURSE VW
THE CUSTARDS
CLARENCE RD
PEMBERTON RD
Custards
Football Grnd
FOREST GDNS
QUEENS PAR
EMPRESS RD
Lyndhurst
KINGS CL
Crown Hotel
WELLANDS
HOMEFIELD ROAD
Fire Sta
Caravan Park
QUEENS RD
PRINCES CRES
SOUTHAMPTON ROAD
A35
Lyndhurst Hill
Swan Green
KNIGHTWOOD AV
KNIGHTWOOD CL
Sch
ELCOMBES CL
HIGH STREET
Liby
CHURCH LA
The Queen's House
LANE
RUFUS CT
BEAULIEU ROAD
B3056
Cemy
Bolton's Bench
Cricket Grnd
The Swan Inn
Cut Walk
ROAD
BOURNEMOUTH ROAD
CHAPEL LANE
HASKELLS CL
SANDY LA
SHRUBBS HILL RD
Social Centre
SHAGGS MDW
DEARING CL
SHRUBBS HILL GDNS
THE GREAT MEAD
LEA CL
MEADOWS
GOSPORT LANE
OAK CL
CEDAR MOUNT
FIR CL
FOX LEASE TER
Appletree Court Council Offices
The Bench
Vernall's Farm
GOOSE GRN
Cuffnell's Farm
Irons Hill Walk
Clayhill
CLAY HILL
Artificial Insemination Centre (Hants Cattle Breeders)
Foxlease Park
BEECHEN LANE
PARK CL
HILLARY CL
Pondhead Inclosure
Crown & Stirrup PH
PINKNEY LANE
Pinkney Farm
Wilverley Park
Swimming Pool
Foxlease House (Girl Guide Association)
Clay Hill

Milford on Sea
Lymore
Lymington Road
Milford Cres
High Street
Keyhaven Road
Sea Road
Hurst Road
Park Lane
Cliff Road
Barnes Lane
Manor Road
Blackbush Road
Lymore Lane
Lymore Valley
Agarton Lane
School Lane
New Lane
Carrington Lane
Knowland Drive
Kivernell Road
De La Warr Road
Shorefield Road
Downton Lane
Dane Road
West Road
Whitby Road
Sharvells Road
Studland Dr
Cox's Bridge
Fish Pond
Barnes Farm
Nursery
Sch
Recreation Ground
Pleasure Grounds
Hospital
St Georges Hospital
Milford Trad Est
Caravan Park
Sturt Pond
Groynes
Shorefield Copse
Blackbush Copse
Studland Common
Danes Stream
Sea Breeze Caravan Site
Shorefield Holiday Park
Shorefield Caravan Park
Holiday Park
Hordle Manor Farm
Hordle House School
Playing Field
Hordle Cliff
Rook Cliff
B3058
Downton
1
2
3
4
A
B
C
D
E
F

1
2
3
4
24
15
F
E
D
C
B
A
Danesbury Farm
Danestream Farm
Stanley's Copse
Stanley Park Caravan & Chalet Site
Bashley Park
Caravan Park
Bashley
Bashley Copse
Great Woar Copse
Pond Copse
School
New Milton
Ashley
Sports Ground
College
Sch
VAGGS LANE
VAGGS LA
SWAY ROAD
B3055
BASHLEY DRIVE
BASHLEY ROAD
B3058
FERNHILL LANE
BASHLEY CROSS ROAD
MARKS LANE
SMITHY LANE
NEW LANE
STEM LANE
BROCKHILLS LANE
ASHLEY COMMON ROAD
ASHLEY ROAD
LOWER ASHLEY ROAD
OAKWOOD AVENUE
OAKWOOD AV
NORTH AVENUE
MANOR ROAD
OSBORNE ROAD
STATION ROAD
HARE LA
ASHLEY LANE
THORNHAM RD
HIGH RIDGE CRESCENT
LYON AVENUE
YORK AVENUE
HERBERT ROAD
SPENCER ROAD
LEIGH ROAD
FERNHILL RD
WHITEFIELD RD
KENNARD ROAD
MARRYAT ROAD
LAKE GROVE ROAD
CULL LANE
CULLWOOD LA
MARSTON ROAD
FERNDALE ROAD
BARRS WOOD RD
MEADOW RD
GAINSBOROUGH
FOREST OAK DR
FOREST PINES
VIOLET LA
BROOK AVENUE
WELLINGTON CT
RUBENS CL
FERNGLADE
UBSDELL CL
WINCHESTER RD
PARK RD
PARK CL
FRAMPTON CL
BURNLEIGH GDNS
HOLLY LA
YEW LA
ACORN CL
PRINCES PL
KINGS RD
LITCHFORD ROAD
WATERFORD RD
HOME WOOD CL
OAK RD
WILLOWDENE CL
QUEENS GRO
BELMONT CL
CUTLER CL
MOLYNEAUX RD
MELROSE CT
ROEBUCK CL
CONWAY CL
STIRLING CL
WARWICK
POWIS CL
KENILWORTH CL
JAUNDRELLS CL
WOODVALE GDNS
MANOR RD
PILGRIMS CL
BERTRAM RD
FRANKLIN RD
MILLER CL
HAWTHORN CL
POPLAR RD
PENNYWELL GDNS
ASHLEY GDNS
SUMMERTREES CT
SUNDEW CL
GORSE RD
COPPICE CL
BLACKTHORN WY
THE MALLOWS
FOXGLOVE PL
VELVET LAWN ROAD
DEER PK CL
FAWN GARDENS
HART CL
DRIVE
THE HYDE
STAG CL
DOE COPSE WAY
ANTLER
DARK LANE
HAZELWOOD AV
ROSEWOOD GDNS
BEECHWOOD
ROSECRAE CL
MARLEY AV
MARLEY
CABOT WY
HOWE CL
HARDY CL
NELSON
DRAKE CL
RALEIGH CL
RAMSEY CT
WALNUT ROAD
KELSALL GDNS
POND CL
CARISBROOKE CT
BROWNSEA CL
CHATSWORTH
LINNET CT
CHAFFINCH CL
DAVIS
WORTHY RD
VINCENT
PLEASANCE WY
HATFIELD CT
WILTON GDNS
BALMORAL WK
BUCKINGHAM WK
STRATFIELD PL
THATCHERS
SHAVES CL
LINHORNS LA
THE LANES
HOLLANDS
WOODLANDS WY
VIXEN WK
HARRIS LA
BADGERS COPSE
PARTRIDGE WY
GRN WOOD
WOTTERS WK
THE FALLOWS
CROCKFORD CL
WHITLEY WY
AKESHILL CL
KESWICK RD
APPLESLADE WY
DERWENT RD
BROOK CT
GRASMERE GDNS
WINTON WY
HILTON RD
GORSEFIELD RD
CHARLES CRES
DEERLEAP WAY
SPINNEY WY
PALMER PL
LINFORD CL
BALLARD CL
LITTLE BARRS DR
BARRS WOOD DR
BARRS AVENUE

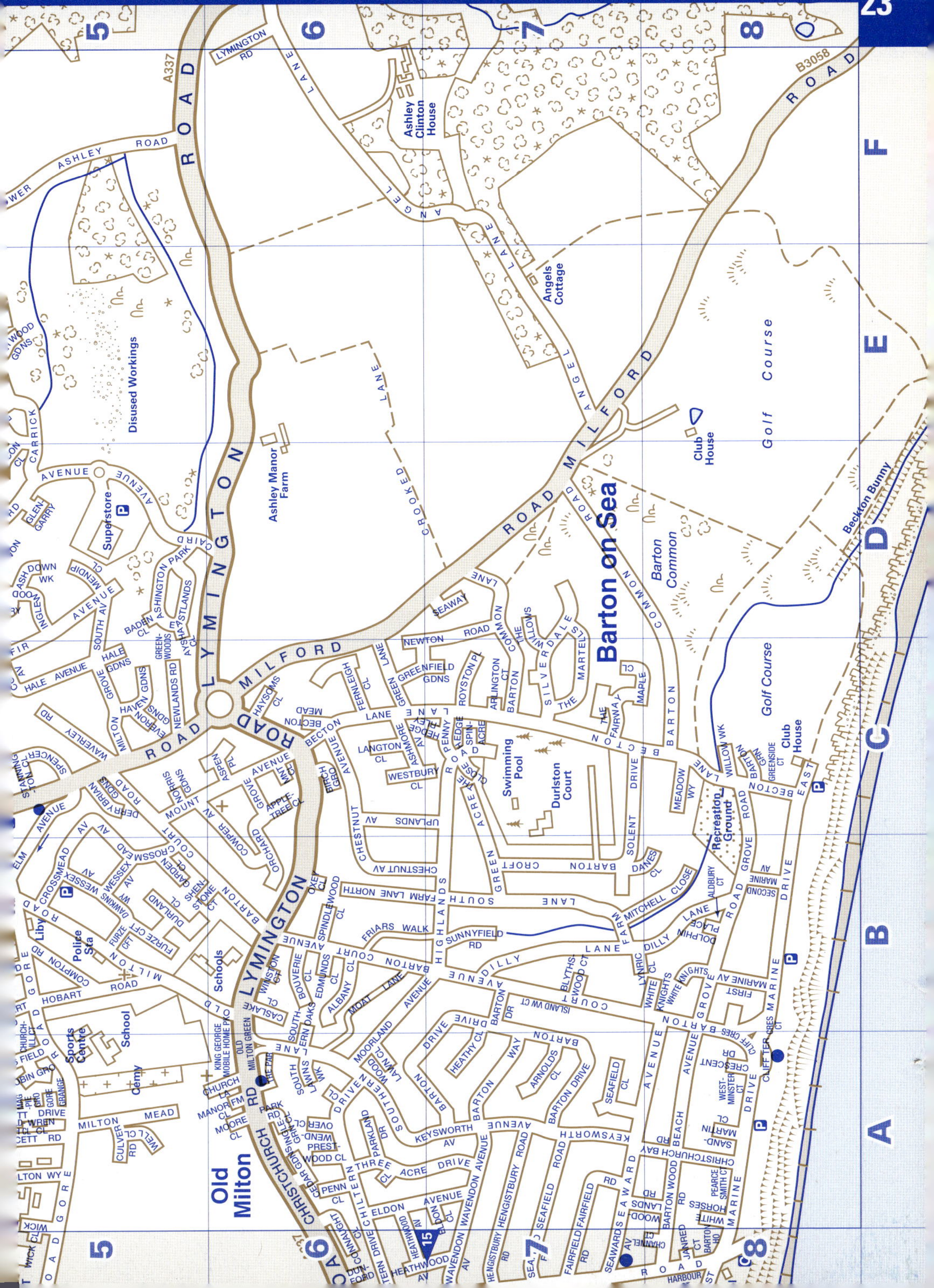

5
6
7
8
F
E
D
C
B
A
A337
B3058
LYMINGTON ROAD
LYMINGTON RD
ASHLEY ROAD
ANGEL LANE
Ashley Clinton House
Angels Cottage
Disused Workings
Ashley Manor Farm
CROOKED LANE
MILFORD ROAD
Club House
Golf Course
Barton on Sea
Barton Common
COMMON ROAD
Beckton Bunny
CARRICK
AVENUE
Superstore
GLEN GARRY
CAIRD
ASHINGTON PARK
ASHDOWN WK
MENDIP CL
INGLEWOOD
SOUTH AV
BADEN CL
EASTLANDS
AYSHA
GREENWOODS
HALE GDNS
HALE AVENUE
GROVE
HAVEN GDNS
NEWLANDS RD
EVERON GDNS
MILTON
WAVERLEY RD
SPENCER
STANNINGTON CL
HAYSOMS CL
BECTON MEAD
BECTON
FERNLEIGH CL
GREEN LANE
NEWTON
GREENFIELD GDNS
ROYSTON PL
SEAWAY
ARLINGTON CT
BARTON COMMON
THE WILLOWS
SILVERDALE
THE MARTELLS
MAPLE CL
THE FAIRWAY
BECTON LANE
BARTON
LANGTON CL
ASHMORE AV
HEDGERLEY
PENNY HEDGE
ACRE
WESTBURY CL
Swimming Pool
Durlston Court
SOLENT DRIVE
MEADOW WY
WILLOW WK
Recreation Ground
BARTON GRN
GREENSIDE CT
BECTON ROAD
EAST
ASPEN PL
MOUNT AVENUE
NORRIS GDNS
DERRYBRIAN GDNS
AVENUE
ELM
GROVE
APPLE TREE CL
MINT
COWPER AV
ORCHARD
BIRCH GRO
UPLANDS AV
CHESTNUT AV
BARTON CROFT
DANES CL
CROSSMEAD AV
WESSEX AV
DAWKINS WY
GARDEN CT
DURLAND
SHENSTONE CT
FURZE CFT
BARTON
COURT
OXEY CL
SPINDLEWOOD CL
FARM LANE NORTH
SOUTH FARM LANE
HIGHLANDS
MITCHELL CLOSE
ALDBURY CT
DOLPHIN PLACE
SECOND MARINE AV
MARINE DRIVE
Police Sta
Liby
Schools
COMPTON RD
HOBART ROAD
MILTON ROAD
FRIARS WALK
SUNNYFIELD RD
DILLY LANE
LYRIC
WHITE KNIGHTS
FIRST MARINE AV
CASLAKE CL
WINSTON CT
BOUVERIE CL
EDMUNDS CL
ALBANY CL
MOAT LANE
AVENUE
BLYTHSWOOD CT
ISLAND VW CT
CLIFF CRES
CLIFFE CT
GORE ROAD
Sports Centre
School
Cemy
KING GEORGE MOBILE HOME PK
OLD MILTON GREEN
THE PAR
SOUTH LAWNS
FERN OAKS
MOORLAND CL
HEATHY CL
BARTON DRIVE
BARTON WAY
ARNOLDS CL
SEAFIELD CL
CRESCENT DR
WESTMINSTER CT
SANDMARTIN CL
CHURCH LA
MANOR FM CL
MOORE CL
MILTON MEAD
WELL CL
CULVER RD
GRANGE
GORE DRIVE
WREN CL
Old Milton
CHRISTCHURCH RD
PARK RD
SINGLETON
OVER CL
WENDOVER
PRESTWOOD CL
PARKLAND DR
SOUTHERN LAWN
KEYSWORTH AV
KEYSWORTH AVENUE
THREE ACRE DRIVE
PENN CL
CEDAR
CHILTERN DRIVE
ELDON AVENUE
ELDON CL
WAVENDON AVENUE
HENGISTBURY ROAD
SEAFIELD ROAD
FAIRFIELD RD
SEAWARD AV
WOODLANDS RD
BARTON WOOD RD
BEACH AVENUE
CHRISTCHURCH BAY RD
PEARCE SMITH CT
HORSES
WHITE BARTON HO
MARINE
CHANNEL CT
JANRED CT
HEATHWOOD AV
DUNFORD
CONNAUGHT
HARBOUR RD
ST
WICK CL
ALTON WY
15

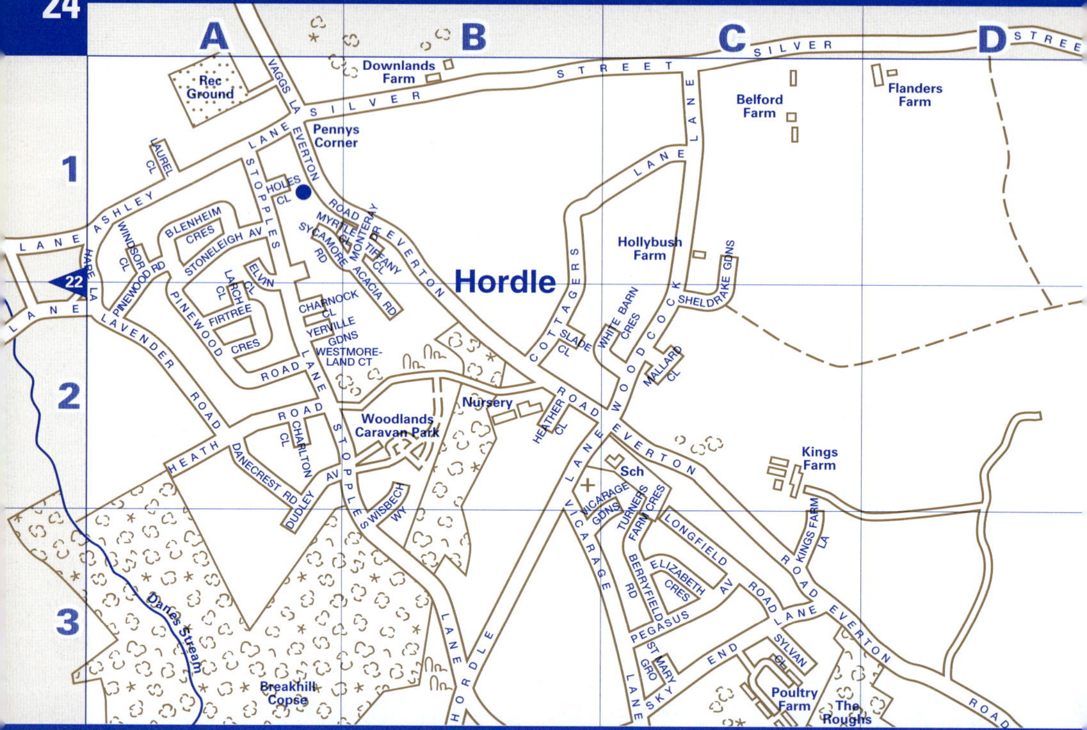
A
B
C
D
1
2
3
Rec Ground
Downlands Farm
VAGGS LA
SILVER STREET
Pennys Corner
LAUREL CL
LANE
EVERTON
STOPPLES
HOLES CL
ROAD
MYRTLE
MONTERAY
TIFFANY CL
SYCAMORE RD
ACACIA RD
ASHLEY LANE
WINDSOR CL
BLENHEIM CRES
STONELEIGH AV
HARE LA
22
PINEWOOD RD
LARCH CL
ELVIN CL
FIRTREE
PINEWOOD CRES
CHARNOCK CL
YERVILLE GDNS
WESTMORE-LAND CT
LAVENDER ROAD
Hordle
COTTAGERS LANE
WHITE BARN CRES
SLADE CL
WOODCOCK LANE
SHELDRAKE GDNS
MALLARD CL
Belford Farm
Flanders Farm
Hollybush Farm
Woodlands Caravan Park
Nursery
HEATHER CL
HEATH ROAD
CHARLTON CL
DANECREST RD
DUDLEY AV
WISBECH WY
Sch
VICARAGE GDNS
TURNERS FARM CRES
LONGFIELD ROAD
VICARAGE LANE
BERRYFIELD RD
ELIZABETH CRES
PEGASUS AV
ST MARY GRO
SKY END LANE
SYLVAN CL
Kings Farm
KINGS FARM LA
Danes Stream
Breakhill Copse
HORDLE LANE
Poultry Farm
The Roughs

EVERTON

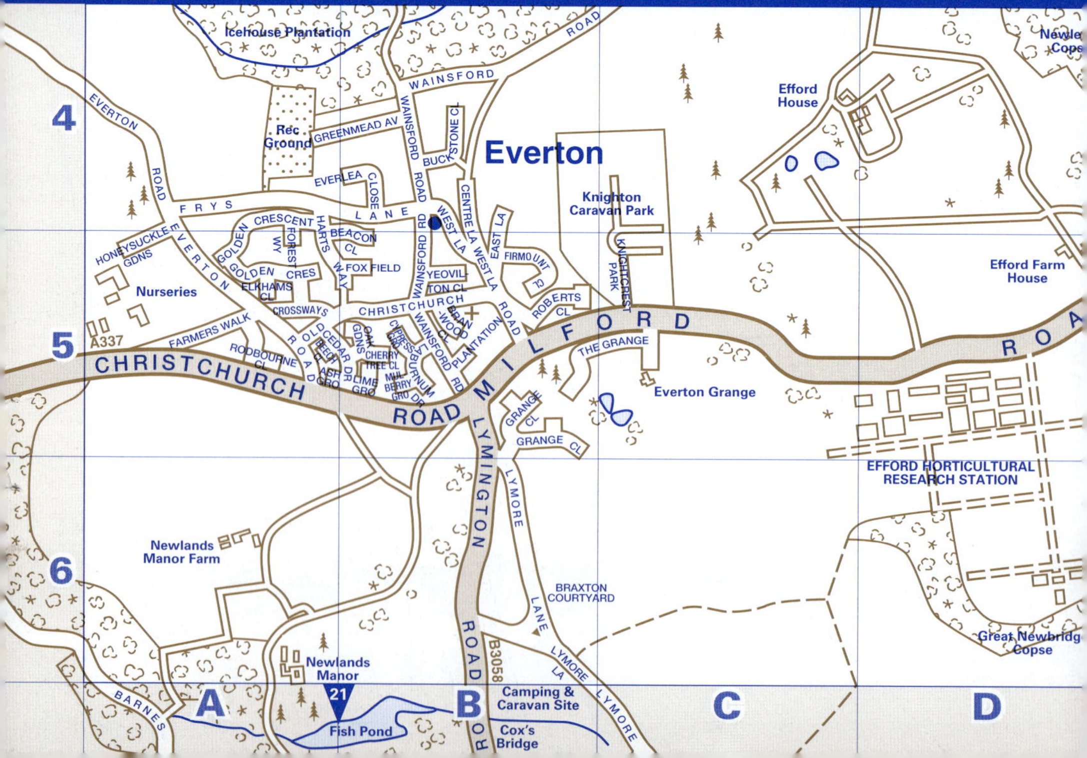
Icehouse Plantation
ROAD
WAINSFORD
4
5
6
EVERTON ROAD
Rec Ground
GREENMEAD AV
WAINSFORD ROAD
STONE CL
BUCKSTONE CL
Everton
EVERLEA CLOSE
FRYS LANE
Knighton Caravan Park
Efford House
Newlands Copse
HONEYSUCKLE GDNS
CRESCENT
GOLDEN CRES
FOREST WY
HARTS WAY
BEACON CL
FOX FIELD
WAINSFORD RD
WEST LA
CENTRE LA
EAST LA
WEST LA
FIRMOUNT CL
YEOVILTON CL
ELKHAMS CL
CROSSWAYS
CHRISTCHURCH ROAD
ROBERTS CL
KNIGHTCREST PARK
Efford Farm House
Nurseries
A337
FARMERS WALK
RODBOURNE CL
OLD ROAD
CEDAR DR
BEECH
ASH GRO
OAK GDNS
CHERRY TREE CL
LIME GRO
MULBERRY GRO
CYPRESS
LABURNUM DR
PLANTATION ROAD
BRANWOOD CL
MILFORD ROAD
THE GRANGE
Everton Grange
GRANGE CL
LYMINGTON ROAD
LYMORE LANE
EFFORD HORTICULTURAL RESEARCH STATION
Newlands Manor Farm
BRAXTON COURTYARD
B3058
Newlands Manor
Great Newbridge Copse
BARNES
21
Fish Pond
Camping & Caravan Site
Cox's Bridge
LYMORE LA
B
C
D

Poulner
Hightown
Ringwood
Whites Copse
Rec Ground
Lynes Farm
Lane End Farm
Hightown Lake
Kingfisher Lake
Headlands Adventure Centre
Playing Fields
Cemy
Sports Ground
Swimming Pool
School
HIGHTOWN IND EST
CROW ARCH LA IND EST
RINGWOOD TRADING ESTATE
MILLSTREAM TRADING EST
CARVERS TRAD EST
Council Offices
Fire Sta
Bickerley Common
Mill Stream
Avon Valley Path
River Avon
Up Mead
Lin Brook
POULNER HILL
RINGWOOD BY-PASS
SALISBURY ROAD
CHRISTCHURCH RD
NOUALE LANE
NARROW LANE
EASTFIELD LANE
SOUTHAMPTON ROAD
NORTH POULNER RD
NORTHFIELD ROAD
HIGHFIELD DRIVE
BROADSHARD LANE
PARSONAGE BARN LANE
KINGSFIELD
HIGHTOWN ROAD
CROW ARCH LANE
CROW LANE
MANSFIELD ROAD
HIGH STREET
A338
B3347
1
2
3
4
A
B
C
D
E
F

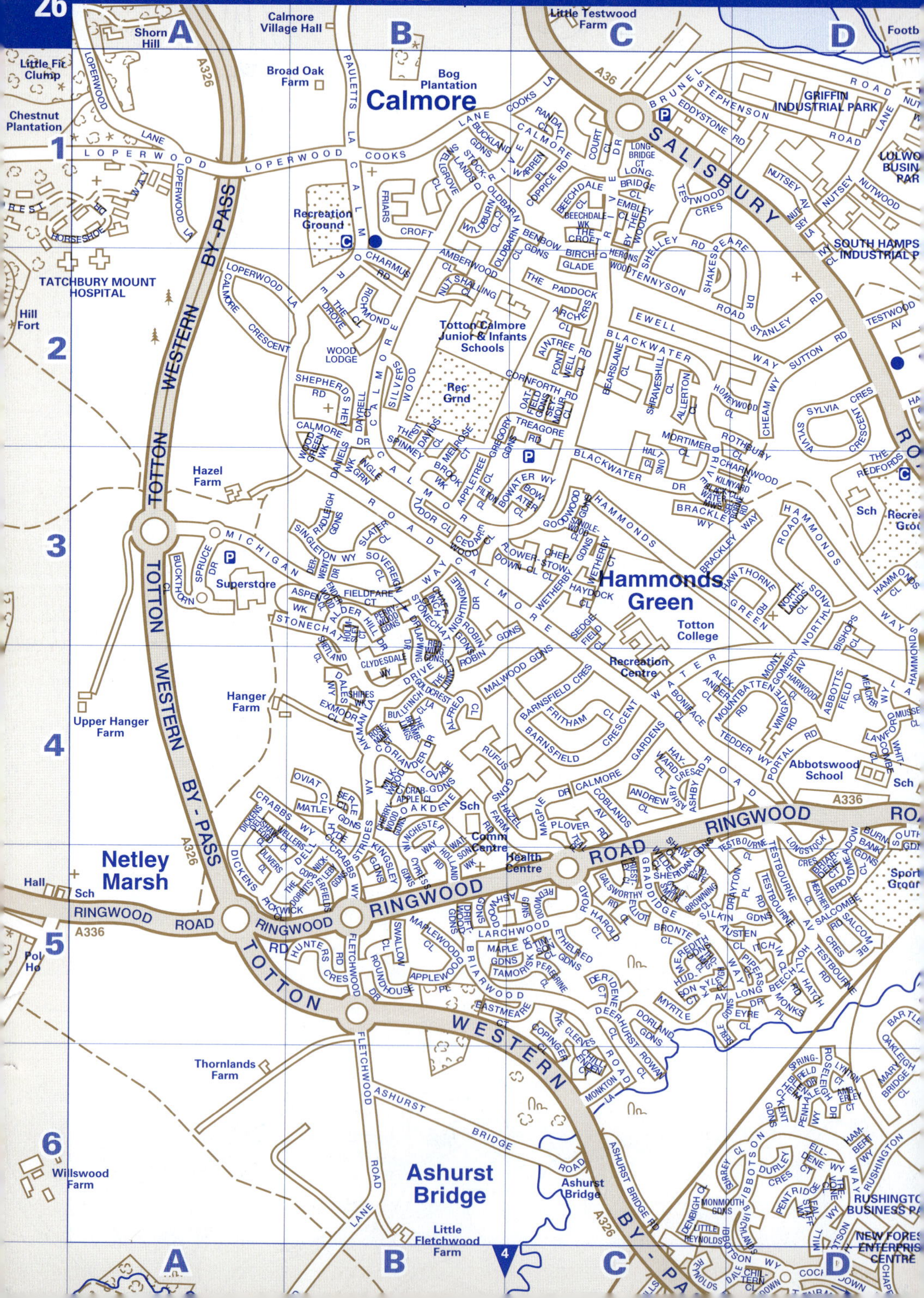
A
B
C
D
1
2
3
4
5
6
Calmore
Hammonds Green
Netley Marsh
Ashurst Bridge
Shorn Hill
Calmore Village Hall
Little Testwood Farm
Little Fit Clump
Chestnut Plantation
Broad Oak Farm
Bog Plantation
GRIFFIN INDUSTRIAL PARK
SOUTH HAMPS INDUSTRIAL P
Recreation Ground
TATCHBURY MOUNT HOSPITAL
Hill Fort
Totton Calmore Junior & Infants Schools
Rec Grnd
Wood Lodge
Hazel Farm
Superstore
Hanger Farm
Upper Hanger Farm
Totton College
Recreation Centre
Abbotswood School
Sch
Comm Centre
Health Centre
Thornlands Farm
Willswood Farm
Little Fletchwood Farm
Hall
A326
A336
A36
TOTTON WESTERN BY-PASS
RINGWOOD ROAD
SALISBURY ROAD
LOPERWOOD LANE
COOKS LANE
CALMORE ROAD
CALMORE DRIVE
CALMORE CRESCENT
MICHIGAN WAY
BLACKWATER DRIVE
HAMMONDS WAY
HAMMONDS GREEN
WATER ROAD
BARNSFIELD CRESCENT
ASHURST BRIDGE ROAD
FLETCHWOOD LANE
FLETCHWOOD ROAD
DEERHURST CRESCENT
TENNYSON ROAD
SHAKESPEARE ROAD
STANLEY ROAD
SUTTON ROAD
EWELL WAY
ASHBY ROAD
BRUNEL ROAD
STEPHENSON ROAD
EDDYSTONE RD
TESTWOOD CRES
TESTWOOD AV
RUSHINGTON BUSINESS PA
NEW FOREST ENTERPRISE CENTRE
4

E
F
G
H
1
2
3
4
5
6
Pumping Station
M271 JUNCTION 1
BROWNHILL WAY
REDBRIDGE LA
LANE
NEW COTTS
TEST
Chadney Meadow
Ruddy Mead
Yarnsey
River Test
TEST VALLEY BUSINESS CENTRE
BROWNHILL ROAD
The Holy Family First & Middle Schools
LOWER
REDBRIDGE
M271
YEOMAN IND PARK
YEOMAN PK
Testwood
Salmon Leap
Test Way
LOWER TEST NATURE RESERVE
Running Track
THOR-NESS CL
PORLOCK RD
VEL CT
DURLSTON RD
Test Playing Field
STUDLAND
Redbridge
BALLARD CL
Playing Field
School
GREENFIELDS AV
GREEN FIELDS CL
CRESCENT
TESTWOOD
ST MICHAELS RD
LANE
STIRLING CL
PEMBROKE CL
STIRLING CRES
WARWICK RD
HURST CL
ROAD
TESTWOOD LA
YORK CL
ARUNDEL RD
GOVER
WESTOVER RD
CONISTON RD
ROAD
THE FURLONGS NATURE RESERVE
REDBRIDGE TOWERS
REDBRIDGE FLYOVER
Redbridge School
SALISBURY
STANNINGTON
CRES
STANNINGTON CRES
Football Ground
TESTWOOD PL
PARKLANDS
Anchor Hotel
Red Bridge
REDBRIDGE CAUSEWAY
CLOVER NOOKE
A33
Youth Club
SUNSET AV
MORPETH AV
Health Centre
Fire Sta
JENNINGS RD
COMPTON RD
OLD REDBRIDGE ROAD
RAILWAY COTTS
TATE RD
PAT BEAR CL
STATION RD
BRUNEL RD
OAK CL
OAKRIDGE RD
MAYFIELD
YEOMANS WY
WESTFIELD
OAK-MOUNT AV
Civic Centre
Amb Sta
CAUSEWAY CRES
Totton
Sch
LEBANON RD
REDBRIDGE
Redbridge Channel
OAKLANDS AV
OAKFIELD RD
Liby
LIBRARY RD
BEAUMONT RD
CAUSEWAY CRES
A36
SALISBURY ROAD ARCADE
COMMERCIAL ROAD
ROAD
HYMAN WAY
BAGBER RD
HASEL-BURY RD
SUNNY WY
POPES LA
Supermarket
GLEN RD
JUNCTION RD
MILL RD
STATION RD
TOTTON
A35
BY-PASS
ROAD
NGWOOD
MAYNARD
ROAD
JUNCTION RD
HIGH ST
TREESIDE AV
Brokenford
BROKENFORD
JACKIE WIGG GDNS
BROKENFORD AV
OSBORNE RD
RUM-BRIDGE GDNS
Works
KINROSS RD
AVENUE
Rushington
CULFORD AV
CUL-FORD WY
RUSHINGTON
RUMBRIDGE ST
WINSOR RD
ROAD
ELING
Depot
Eling Great Marsh
River Test
EVERGREENS
FISHERS RD
ROSE RD
Sch
KINROSS AV
BARTRAM RD
SCHOOL RD
Eling Wharf
FRAMPTON WAY
BY-PASS
Rum Bridge
Recreation Ground
AVENUE
Works
East Quay
LANE
HILL
MARCHWOOD
PARK RD
DOWNS
LEXBY RD
BOOTHBY CL
Mus
SPICERS
COPSE CL
THE ROBERTS RD
DRIVE
MILVERTON RD
MILV-ERTON CL
DOWNS PARK
PARK CRES
Tide Mill
Toll Bridge
Eling Channel
ELING
PARKSIDE
Bartley Water
Eling Hill
HOUNS-DOWN CL
POWELL CRES
HOUNSDOWN AV
Cemy
Eling
MEADOW CL
ORCHARD CL
BROOKSIDE
BY-PASS
HILL
Village Bells P.H

A - Z INDEX TO STREETS
with Postcodes

The Index includes some names for which there is insufficient space on the maps. These names are indicated by an * and are followed by the nearest adjoining thoroughfare.

A Avenue SO45 10 A1
Abbey Cl SO45 17 F3
Abbots Cl BH23 14 C5
Abbotsford SO40 9 D3
Abbottsfield SO40 26 D4
Abingdon Dr BH23 15 F4
Acacia Rd SO41 24 B1
Acorn Cl BH25 22 D3
Adams Rd SO45 17 F4
Addington Ct SO41 21 D2
Addington Pl BH23 12 D4
Addiscombe Rd BH23 12 A3
Addison Rd SO42 7 C5
Addison Sq BH24 25 C3
Adlams La SO41 6 A4
Admirals Cl SO45 10 D2
Admirals Way SO45 17 G2
Agarton La SO41 21 F1
Aikman La SO40 26 B4
Aintree Rd SO40 26 C2
Airfield Rd BH23 13 F4
Airfield Way BH23 13 F3
Airspeed Rd BH23 13 G3
Akeshill Cl BH25 22 C2
Alan Ct BH23 15 E5
Albany Cl BH25 23 B6
Albert Rd BH25 22 B4
Albion Rd SP6 11 F2
Aldbury Ct BH25 23 B8
Alder Cl, Christchurch BH23 12 C1
Alder Cl, Southampton SO45 16 C4
Alder Hill Dr SO40 26 B3
Alexander Cl, Southampton SO40 26 C4
Alexander Cl, Christchurch BH23 13 E4
Alexandra Cl SO45 17 F3
Alexandra Rd, Lymington SO41 18 D2
Alexandra Rd, Fordingbridge SP6 11 E3
Alexandra Rd, Southampton SO45 17 F2
Alfred Cl SO40 26 B4
All Saints Rd SO41 19 F5
Allenwater Dr SP6 11 D2
Allerton Cl SO40 26 C2
Ambassador Cl BH23 13 G4
Amberley Cl BH23 14 C4
Amberley Ct SO40 26 D6
Amberslade Walk SO45 17 E5
Amberwood Cl, Southampton SO40 26 B2
Amberwood Cl, Christchurch BH23 14 D3
Amberwood Dr BH23 14 D2
Amberwood Gdns BH23 14 D2
Ambleside Rd SO41 19 F4
Ambury La BH23 13 E2
Amethyst Rd BH23 13 E3
Amsterdam Sq BH23 12 C4
Anchor Cl BH23 13 G6
Anchor Mews SO41 19 F3
Anchorage Way SO41 19 E4
Anderwood Dr SO41 6 B5
Andover Cl BH23 13 H3
Andrew Cl, Southampton SO40 26 C4
Andrew Cl, Hythe SO45 17 F6
Andrew La BH25 22 E4
Androse Gdns BH24 25 B4
Angel La BH25 23 E7
Angeline Cl BH23 14 D4
Anne Cl BH23 12 A1
Anson Cl, Ringwood BH24 25 E2
Anson Cl, Christchurch BH23 13 F4
Antler Dr BH25 22 A3
Appleslade Way BH25 22 C2
Appletree Cl, Southampton SO40 26 B3
Appletree Cl, New Milton BH25 23 C6
Applewood Pl SO40 26 B5
Archers Cl SO40 26 C2
Archgate SO41 19 E2
Arden Walk BH25 22 D4
Argyle Rd BH23 13 E6
Ariel Cl BH6 12 A6
Ariel Dr BH6 12 B6
Arlington Ct BH25 23 C7
Armada Dr SO45 17 E4
Armitage Av SO45 17 E6
Armstrong Cl SO42 7 B3
Armstrong La SO42 7 B3
Armstrong Rd SO42 7 B3
Arnolds Cl BH25 23 A7
Arnwood Av SO45 17 E6
Arran Way BH23 15 F3
Arthur La BH23 12 A3
Arthur Rd BH23 12 A3
Arundel Cl BH25 15 H1
Arundel Rd SO40 27 F3
Arundel Way BH23 14 C5
Ash Gro, Lymington SO41 24 A5
Ash Gro, Southampton SO40 4 B3
Ash Gro, Ringwood BH24 25 D3
Ash Rd SO40 4 A4
Ashburn Garth BH24 25 E4
Ashburton Cl SO45 16 C3
Ashby Cres SO40 26 C4
Ashby Rd SO40 26 C4
Ashdene Rd SO40 4 B3
Ashdown SO45 10 B2
Ashdown Rd SO45 10 B2
Ashdown Walk BH25 23 D5
Ashford Cl SP6 11 C3
Ashford Cres SO45 17 G3
Ashford Rd SP6 11 B4
Ashington Pk BH25 23 D5
Ashleigh Cl SO45 17 F6
Ashlet Gdns BH25 22 E3
Ashlett Cl SO45 10 D2
Ashlett Mews SO45 10 D2
Ashlett Rd SO45 10 D2
Ashley Arnewood Ct*, Ashley Rd BH25 22 D4
Ashley Cl BH24 25 E4
Ashley Common Rd BH25 22 D2
Ashley La, Hordle SO41 24 A1
Ashley La, New Milton BH25 22 E3
Ashley La, Lymington SO41 19 F3
Ashley Meads BH25 22 E3
Ashley Rd BH25 22 C4
Ashmore Av BH25 23 C6
Ashmore Gro BH23 14 C3
Ashtree Cl BH25 22 E4
Ashurst Bridge Rd SO40 26 B6
Ashurst Cl SO40 4 B3
Ashwood Gdns SO40 26 C5
Aspen Pl BH25 23 C6
Aspen Walk SO40 26 B3
Asquith Cl BH23 12 D5
Astra Ct SO45 17 F1
Atheling Rd SO45 17 F3
Aubrey Cl SO41 21 F3
Auckland Av SO42 7 C4
Auckland Pl SO42 7 C4
Auckland Rd BH23 14 A5
Audemer Ct BH24 25 E2
Austen Cl SO40 26 C5
Auster Cl BH23 13 G3
Autumn Copse BH25 22 E4
Avenue Rd, New Milton BH25 22 B4
Avenue Rd, Brockenhurst SO42 7 C4
Avenue Rd, Lymington SO41 19 E3
Avenue Rd, Christchurch BH23 15 F3
Avon Bldgs BH23 12 B3
Avon Cl SO41 18 D3
Avon Meade SP6 11 D2
Avon Rd East BH23 12 A2
Avon Run Cl BH23 13 G5
Avon Run Rd BH23 13 H5
Avon Wharf BH23 12 C4
Aysha Cl BH25 23 C5

B Avenue SO45 10 A1
Back La, Lymington SO41 6 C5
Back La, Brockenhurst SO42 7 D4
Back La, Ringwood BH24 25 F3
Baden Cl BH25 23 D5
Badgers Cl SO41 6 C5
Badgers Copse BH25 22 D1
Badgers Walk SO45 17 E5
Badminston Drove SO45 10 D4
Bagber Rd SO40 27 E4
Bailey Cl BH25 22 E3
Baldwin Cl BH23 12 D4
Balfour Cl BH23 14 A4
Ballard Cl, New Milton BH25 22 C3
Ballard Cl, Southampton SO16 27 H3
Balmer Lawn Rd SO42 7 D1
Balmoral Ct BH23 15 E5
Balmoral Walk BH25 22 A4
Bank Cl BH23 12 B4
Bankhill Dr SO41 19 E1
Bankview SO41 19 E1
Barclay Mews SO45 17 F6
Bargates BH23 12 A3
Baring Rd BH6 12 A6
Barleycorn Walk SO40 9 B2
Barnes La SO41 21 D1
Barney Hayes La SO40 9 D2
Barnfield BH23 14 A5
Barnsfield Cres SO40 26 C4
Barrs Av BH25 22 C3
Barrs Wood Dr BH25 22 C3
Barrs Wood Rd BH25 22 D3
Bartley Av SO40 26 D5
Barton Cft BH25 23 B7
Barton Common La BH25 23 C7
Barton Common Rd BH25 23 C8
Barton Court Av BH25 23 B8
Barton Court Rd BH25 23 B6
Barton Dr BH25 23 A7
Barton Grn BH25 23 C8
Barton Ho BH25 15 H6
Barton La BH25 15 H4
Barton Way BH25 23 A7
Barton Wood Rd BH25 23 A8
Bartons Rd SP6 11 E3
Bartonside Rd BH25 15 F5
Bartram Rd SO40 27 F5
Bashley Cross Rd BH25 22 A2
Bashley Dr BH25 22 C1
Bashley Rd BH25 22 B1
Bath Rd SO41 19 G4
Batten Cl BH23 12 D3
Bay Tree Way BH23 14 B3
Bays Rd SO41 18 C3
Beach Av BH25 23 A8
Beacon Cl SO41 24 A5
Beacon Ct, Christchurch BH23 14 D5
Beacon Ct, Fordingbridge SP6 11 E2
Beacon Dr BH23 14 D5
Beaconsfield Rd BH23 12 B3
Bearslane Cl SO40 26 C2
Beatty Cl BH24 25 E2
Beaufort Cl BH23 13 H3
Beaulieu Cl BH25 15 H1
Beaulieu Rd, Lynhurst SO43 20 C4
Beaulieu Rd, Southampton SO45 16 D6
Beaumaris Gdns SO45 17 E2
Beaumont Rd SO40 27 F4
Beckley Copse BH23 15 E3
Becton La BH25 23 C8
Becton Mead BH25 23 C6
Bedford Cl SP6 11 E2
Beech Cl SO41 24 A5
Beech Rd SO40 4 B3
Beechcroft La BH24 25 C2
Beechdale Cl SO40 26 C1
Beechdale Walk SO40 26 C1
Beechen La SO43 20 C5
Beechwood SP6 11 D4
Beechwood Av BH25 22 A3
Beechwood La BH24 8 D2
Beechwood Rd SO40 9 B4
Beechwood Way SO45 16 C4
Belfield Rd BH6 12 A6
Bell Cl SO45 10 A3
Bellflower Cl BH23 13 G2
Belmont Cl SO45 17 F5
Belmont Rd BH25 22 E3
Belmore La SO41 19 E4
Belmore Rd SO41 19 E4
Belstone Rd SO40 27 E4
Belvedere Rd, Christchurch BH23 12 A3
Belvedere Rd, Southampton SO45 17 F5
Benbow Gdns SO40 26 C1
Benmore Cl BH25 23 E5
Bennetts La BH24 8 F3
Benson Cl BH23 6 C2
Beresford Gdns BH23 13 E4
Beresford Rd SO41 18 C3
Bermuda Ct BH23 15 E5
Bernwood Gro SO45 10 A6
Berryfield Rd SO41 24 C3
Bertram Rd BH25 22 D3
Betsy Cl BH23 6 C1
Betsy La BH23 6 C1
Beverley Rd SO45 17 E6
Bevis Cl SO45 10 B3
Bickerley Gdns BH24 25 B4
Bickerley Rd BH24 25 A3
Bickerley Ter BH24 25 A3
Billington Pl SO41 18 D5
Bingham Cl BH23 13 E3
Bingham Dr SO41 19 F4
Bingham Rd BH23 12 D3
Birch Gro BH25 23 C6
Birchdale SO45 17 G5
Birchglade SO40 26 C2
Birchlands SO40 26 D6
Birchwood Cl BH23 14 C4
Birchy Hill SO41 6 C6
Bishop Ct BH24 25 C3
Bishops Cl SO40 26 D3
Bisterne Cl, Ringwood BH24 8 F3
Bisterne Cl, Ringwood BH24 8 F4
Bitterne Way SO41 19 E5
Blackberry La BH23 13 E4
Blackbird Way BH23 6 D2
Blackbush Rd SO41 21 C2
Blackdown Cl SO45 16 D4
Blackfield Rd, Southampton SO45 10 A4
Blackfield Rd, Southampton SO45 10 B3
Blackthorn Cl SO41 18 B5
Blackthorn Way BH25 22 E3
Blackwater Dr SO40 26 C2
Blackwater Mews SO40 26 C3
Blair Cl BH25 22 A4
Blandford Ct SO41 21 D2
Blenheim Cl SO40 26 D6
Blenheim Cres SO41 24 A1
Blenheim Dr BH23 13 G4
Blenheim Gdns SO45 16 C5
Blenheim Mews SO45 17 F6
Bluebell Cl BH23 13 H2
Blythswood Ct BH25 23 B7
Boakes Pl SO40 4 B3
Bockhampton Rd BH23 6 B3
Bodowen Rd BH23 12 D1
Boldre Cl BH25 15 G5
Bond Cl SO41 6
Boniface Cl SO40 26
Bonington Cl BH23 13
Boothby Cl SO40 27
Boundstone SO45 17
Bourne La SO40 9
Bourne Rd SO40 9
Bournemouth Rd SO43 20
Bouverie Cl BH25 23
Bowater Cl SO40 26
Bowater Way SO40 26
Bowerwood Rd SP6 11
Bowland Rise BH25 22
Bowling Green Rd SO41 18
Brabazon Dr BH23 13
Bracken Way BH23 15
Brackens Way SO41 19
Brackley Way SO40 26
Braehead SO45 17
Braemar Av BH6 12
Braemar Cl BH6 12
Braemar Dr BH23 14
Bramble Ct*, Fairview Par SO45 17
Bramble La BH23 15
Bramble Walk SO41 18
Bramble Way BH23 6
Bramley Cl SO41 19
Bramshaw Way BH25 15
Bramshott Hill SO45 16
Branders Cl BH6 12
Branders La BH6 12
Branksome Cl BH25 22
Bransgore Gdns BH23 6
Branwell Cl BH23 12
Branwood Cl SO41 24
Braxton Courtyard SO41 24
Breamore Cl BH25 22
Brecon Cl, Southampton SO45 16
Brecon Cl, New Milton BH25 23
Brendon Cl SO45 16
Briar Cl BH23 13
Briar Ct SO45 17
Briardene Ct SO40 26
Briarswood Rise SO45 16
Briarwood Rd SO40 26
Brickfield La SO41 19
Bridge Rd SO41 19
Bridge St, Christchurch BH23 12
Bridge St, Fordingbridge SP6 11
Brighton Rd SO41 6
Brinton La SO45 17
Britannia Way BH23 13
Broad La SO41 19
Broadfields Cl SO41 21
Broadlands Cl BH23 15
Broadlands Rd SO42 7
Broadly Cl SO41 18
Broadmead Cl SO41 19
Broadmeadow Cl SO40 26
Broadshard Ct BH24 25
Broadshard La BH24 25
Broadway BH6 12
Brockhills La BH25 22
Brockishill Rd SO40 9
Brocks Cl SO45 16
Brokenford Av SO40 27
Brokenford La SO40 27
Bronte Av BH23 12
Bronte Cl SO40 26
Brook Av BH25 22
Brook Av North BH25 22
Brook La BH23 6
Brook Rd SO41 19
Brook Ter SP6 11
Brook Walk SO40 26
Brook Way BH23 13
Brookland Cl SO41 18
Brookley Rd SO42 7
Brooklyn Ct BH25 22
Brooks Cl BH24 25
Brookside SO40 27
Brookside Cl BH23 6

rookside Rd, Christchurch BH23 6 B1
rookside Rd, Brockenhurst SO42 7 C3
rookside Way BH23 14 C3
roomfield La SO41 19 E2
roomhill Cl SO41 18 B5
roomy Cl SO45 16 C3
roughton Rd SO43 20 B2
rownhill Way SO16 27 H1
rowning Cl SO40 26 C5
rownings Cl SO41 18 B3
rownsea Cl BH25 22 A4
runel Rd, Southampton SO40 26 C1
unel Rd, Totton SO40 27 E1
unel Rd, Redbridge SO15 27 H4
unswick Pl SO41 19 E3
uyn Ct SP6 11 F3
uyn Rd SP6 11 F3
ympton Cl SP6 11 C3
ub La BH23 13 E4
uccaneers Cl BH23 12 D4
ucehayes Cl BH23 14 D5
uchan Ct SO45 16 C4
uckingham Ct BH23 15 E5
uckingham Walk BH25 22 A4
uckland Dene SO41 19 E1
uckland Gdns SO40 26 B1
uckland Gro BH23 14 B3
uckstone Cl SO41 24 B4
uckthorn Cl SO40 26 A3
uldowne Walk SO41 6 B4
ullfinch Cl SO40 26 B4
ullrush Cl SO45 17 E5
urdock Cl BH23 13 H1
ure Cl BH23 13 H4
ure Haven Dr BH23 13 G5
ure Homage Gdns BH23 13 G4
ure Homage La BH23 13 G4
ure La BH23 13 G5
ure Pk BH23 13 H4
ure Rd BH23 13 H4
urford La SO42 7 D2
urgate Flds SP6 11 F2
urley Cl BH25 15 G4
urley Lawn BH24 8 D2
urley Rd, Christchurch BH23 6 A3
rley Rd, Brockenhurst SO42 7 A4
rnbank Gdns SO40 26 D4
rnett Cl SO45 17 G4
rnham Rd SP6 11 E2
rnleigh Gdns BH25 22 E3
rnside BH23 14 B5
rnt House La BH23 6 B1
rrard Gro SO41 19 F5
rton Cl BH23 12 C1
rton Rd BH23 12 D3
shells Farm SP6 11 E4
te Dr BH23 15 F5
tlers La BH24 25 E1
tercup Cl SO45 17 F5
tercup Dr BH23 13 H1
ts Ash Av SO45 17 F6
ts Ash Gdns SO45 17 F6
ts Ash La SO45 17 E6
ts Bridge Hill SO45 17 F4
ts Bridge Rd SO45 17 F5
ts Lawn SO42 7 C2
The Wood SO40 26 C1
ways SO45 17 E4
Pass Rd SO40 27 E5
on Rd BH25 15 H4

venue SO45 10 A1
oot Dr SO45 16 C3
oot Way BH25 22 A4
lhay Cl BH25 22 A4
lnam La SO40 9 A1
logan Rd BH24 25 C3
rd Av BH25 22 D4
edonia Dr SO45 16 C3
edonian Cl BH23 13 H3
kin Cl BH23 12 A1
nore Cres SO40 26 A2
nore Dr SO40 26 B2
nore Gdns SO40 26 C4
nore Rd SO40 26 B1
oe Av SO43 20 B3
shot Rd SO45 10 D2
nbria Dr SO45 16 C3
Camden Hurst SO41 21 B3
Camellia Gdns BH25 22 C4
Cameron Rd BH23 12 D3
Campion Gro BH23 13 E4
Campion Way SO41 19 E1
Cannon St SO41 19 F3
Canons Walk SO41 21 E2
Canterbury Dr SO45 16 C3
Canute Dr BH23 6 C1
Capella Gdns SO45 16 D3
Capesthorne BH23 13 G5
Captains Row SO41 19 F3
Careys Cotts SO42 7 C2
Carisbrooke Ct, New Milton BH25 22 A4
Carisbrooke Ct, Christchurch BH23 15 E5
Carisbrooke Way BH23 14 C4
Carlton Av BH25 15 H5
Caroline Av BH23 13 E5
Carpenter Cl, Lymington SO41 19 E2
Carpenter Cl, Southampton SO45 17 F3
Carradale BH23 13 H2
Carrick Way BH25 23 D5
Carrington Cl SO41 21 F3
Carrington La SO41 21 F3
Carvers La BH24 25 B3
Caslake Cl BH25 23 B6
Castle Av BH23 14 C5
Castle Cl SO41 21 E4
Castle Ct BH23 15 E5
Castle Hill La BH24 8 A2
Castle St BH23 12 B4
Castleman Way BH24 25 B4
Catalina Cl BH23 13 G4
Cathay Gdns SO45 16 C3
Causeway Cres SO40 27 F4
Cavalier Cl SO45 16 C3
Cavendish Corner Caravan Pk BH24 25 D2
Caxton Cl BH23 13 F3
Cecil Av SO40 4 B3
Cedar Dr SO41 24 A5
Cedar Gdns BH25 23 A6
Cedar Mount SO43 20 B4
Cedar Pl BH23 6 C1
Cedarwood Cl SO40 26 B3
Cedric Cl SO45 10 B5
Celandine Cl BH23 13 H2
Centenary Cl SO41 6 C5
Centenary Ho BH23 12 A3
Centre La SO41 24 B4
Centre Pl BH24 25 A3
Chaffey Cl BH24 25 E2
Chaffinch Cl, New Milton BH25 22 A4
Chaffinch Cl, Southampton SO40 26 B3
Chalewood Rd SO45 10 A6
Challenger Way SO45 16 C4
Chaloner Cres SO45 17 F6
Champion Cl SO41 21 F3
Channel Ct BH25 15 H5
Chant Cl BH23 12 D3
Chantry Cl BH23 14 C3
Chapel La, Blackfield SO45 10 A6
Chapel La, Christchurch BH23 6 B2
Chapel La, Fawley SO45 10 B2
Chapel La, Lynhurst SO43 20 B4
Chapel La, Ringwood BH24 8 C3
Chapel La, Southampton SO40 4 D1
Chapel La, Lymington SO41 6 D6
Charing Cl BH24 25 C4
Charles Cres BH25 22 D2
Charles Ley Ct SO45 10 C2
Charles Rd BH23 13 F2
Charleston Rd SO45 17 H6
Charlotte Cl BH23 13 F4
Charlton Cl SO41 24 A2
Charmus Rd SO40 26 B2
Charnock Cl SO41 24 A2
Charnwood Cl SO40 26 C3
Charnwood Dr SP6 11 E2
Chatsworth Way BH25 15 H1
Chaucer Dr SO41 21 D3
Chaveney Cl SO45 17 E6
Cheam Way SO40 26 D2
Chepstow Cl SO40 26 C3
Chequers Cl SO41 18 C4
Cherry Tree Cl SO41 24 B5
Cherry Tree Ct*, Station Rd BH25 23 C5
Cherrywood Gdns SO40 26 B4
Chestnut Av BH25 23 B7
Chestnut Dr SO40 4 C3
Chestnut Rd SO42 7 D3
Cheviot Ct BH23 13 E2
Cheviot Dr SO45 16 C3
Chewton Common Rd BH23 14 D3
Chewton Farm Rd BH23 15 F4
Chewton Lodge BH23 15 F4
Chewton Way BH23 15 E3
Chichester Rd BH24 25 E2
Chichester Way BH23 13 G6
Chillenden Ct SO40 26 C6
Chiltern Cl, New Milton BH25 23 A6
Chiltern Cl, Southampton SO40 4 D1
Chiltern Ct*, Hunt Rd BH23 13 F2
Chiltern Dr BH25 15 H4
Chinham Rd SO40 9 D3
Christchurch Bay Rd BH25 23 A8
Christchurch By-Pass BH23 12 B3
Christchurch Rd, Lymington SO41 24 A5
Christchurch Rd, New Milton BH25 23 A6
Christchurch Rd, Ringwood BH24 25 B3
Christchurch Rd, Christchurch BH23 15 F4
Church Farm SP6 11 E4
Church Hill SO41 21 E3
Church La, New Milton BH25 23 A5
Church La, Southampton SO45 10 B2
Church La, Christchurch BH23 12 B4
Church La, Lynhurst SO43 20 C4
Church La, Sway SO41 6 C6
Church La, Ringwood BH24 8 D3
Church La, Brockenhurst SO42 7 D4
Church La, Lymington SO41 19 E3
Church Mead SO41 19 F5
Church St, Christchurch BH23 12 B4
Church St, Fordingbridge SP6 11 E4
Churchfields SO45 10 D1
Churchill Ct BH25 23 A5
Claire Ct BH23 14 D5
Clare Gdns SO45 10 A6
Clare Lodge Cl BH23 6 B2
Clarence Rd SO43 20 C3
Clarendon Pk SO41 18 D5
Clarendon Rd BH23 12 A3
Clarks Cl BH24 25 B3
Clausen Way SO41 18 D6
Clay Hill SO43 20 C5
Claypits La SO45 16 B4
Clematis Cl BH23 13 H2
Cleveland Cl BH25 15 H5
Cleveland Dr SO45 16 D4
Cliff Cres BH25 23 A8
Cliff Dr BH23 13 H4
Cliff Rd SO41 21 A2
Cliff Ter BH25 23 A8
Cliffe Rd BH25 15 H5
Clinton Cl BH23 15 E3
Clinton Rd SO41 19 E2
Clive Rd BH23 14 B3
Clough La BH24 8 C3
Cloughs Rd BH24 25 C3
Clover Cl BH23 13 H2
Clover Ct BH25 22 E3
Clover Nooke SO15 27 H3
Clydesdale Way SO40 26 B4
Coastguard Cotts BH25 15 H5
Coastguard Way BH23 13 E6
Coat Gdns SO45 17 F3
Coblands Av SO40 26 C4
Cocklydown La SO40 4 D1
Colborne Cl SO41 19 E1
Colbourne Cl BH23 6 B2
Colemere Gdns BH23 14 C4
Coleridge Grn BH23 13 F3
Coleville Av SO45 10 C2
College Rd BH24 25 B3
Collins La BH24 25 B3
Collyers Rd SO42 7 C5
Columbine Cl BH23 13 H1
Comet Way BH23 13 G4
Commercial Rd SO40 27 F4
Compton Rd, New Milton BH25 23 B5
Compton Rd, Southampton SO40 27 F4
Conference Pl SO41 19 F5
Conifer Cl SO45 17 E3
Conifer Cres SO41 18 C4
Coniston Rd, Ringwood BH24 25 C4
Coniston Rd, Southampton SO16 27 H3
Connaught Cl BH25 15 H4
Conway Cl BH25 22 D4
Cook Cl BH24 25 E2
Cooks La SO40 26 B1
Coombe La SO41 6 D5
Cooper Rd SO40 4 B3
Copinger Cl SO40 26 C5
Copperfields SO40 26 B5
Coppice Cl BH25 22 E3
Coppice Rd SO40 26 C1
Copse Av BH25 23 C5
Copse Cl, Hythe SO45 17 E5
Copse Cl, Southampton SO40 27 E6
Copse Rd, New Milton BH25 22 C4
Copse Rd, Ringwood BH24 8 C4
Copse Way BH23 14 B4
Copsewood Rd, Ashurst SO40 4 B2
Copsewood Rd, Hythe SO45 17 E3
Copthorne La SO45 10 D2
Corbin Rd SO41 18 B4
Corbould Rd SO45 17 E6
Cordelia Cl SO45 16 D3
Coriander Dr SO40 26 B4
Corinna Gdns SO45 16 D3
Cormorant Dr SO45 17 H4
Cornflower Dr BH23 13 H1
Cornford Way BH23 14 B5
Cornforth Rd SO40 26 B2
Cornwallis Rd SO41 21 C3
Corsair Dr SO45 16 D3
Cosworth Dr SO45 16 C4
Cotswold Cl SO45 16 D4
Cotswold Ct*, Hunt Rd BH23 13 F2
Cott La BH24 8 D3
Cottage Mews SP6 11 E3
Cottagers La SO41 24 B2
Court Cl, Southampton SO40 26 C1
Court Cl, Christchurch BH23 12 D4
Court Cl, Lymington SO41 19 E5
Courtenay Pl SO41 19 E4
Courtier Cl SO45 16 C3
Cowley Rd SO41 18 C3
Cowper Av BH25 23 B6
Coxstone La BH24 25 B4
Crabapple Cl SO40 26 B4
Crabbs Way SO40 26 A4
Cranborne Pl BH25 22 A4
Cranemoor Av BH23 14 C3
Cranemoor Cl BH23 14 C2
Cranemoor Gdns BH23 14 C3
Cranwell Cl BH23 6 C2
Creedy Path BH23 12 B4
Crescent Ct BH25 23 A8
Crescent Dr BH25 23 A8
Crete Cotts SO45 17 E6
Crete La SO45 17 E6
Crete Rd SO45 17 E6
Cricket Cl BH23 13 F5
Cringle Av BH6 12 A6
Crispin Cl BH23 14 D4
Crittall Cl SO41 6 C5
Crockford Cl BH25 22 C2
Croft Rd, Ringwood BH24 25 E1
Croft Rd, Christchurch BH23 13 F3
Cromalt Cl SO45 16 C4
Crooked La BH25 23 D7
Crossmead Av BH25 23 B5
Crossways SO41 24 A5
Crow Arch La BH24 25 C4
Crow La BH24 25 D4
Cruse Cl SO41 6 B5
Cuckmere La SO16 27 H3
Cuckoo Hill Way BH23 6 D2
Culford Av SO40 27 E5
Culford Way SO40 27 E5
Cull La BH25 22 C1
Cullwood La BH25 22 D2
Culver Rd BH25 23 A5
Culverley Cl SO42 7 C4
Cumberland Way SO45 16 C3
Cunningham Cl, Ringwood BH24 25 E2
Cunningham Cl, Christchurch BH23 13 F4
Cunningham Ct*, Nelson Cl BH25 22 A4
Curlew Cl SO45 17 H5
Curlew Dr SO45 17 G4
Curlew Rd BH23 13 F4
Curlew Walk SO45 17 G4
Curzon Pl SO41 18 D5
Curzon Way BH23 14 B5
Cutler Cl BH25 22 E3
Cygnus Gdns SO45 16 B3
Cypress Gdns SO40 26 B5
Cypress Gro SO41 24 B5
Cyprus Av SO40 4 C3

D Avenue SO45 10 C1
Dacres Walk SO41 21 E2
Dairy Cl BH23 12 D4
Dakota Cl BH23 13 G3
Dale Rd SO45 17 E3
Dales Way SO40 26 B4
Dane Cl SO45 10 B4
Dane Rd SO41 21 A1
Danecrest Rd SO41 24 A2
Danes Cl BH25 23 B7
Danestream Cl SO41 21 D3
Danestream Ct SO41 21 E3
Daneswood Rd BH25 22 D4
Daniells Cl SO41 19 F4
Daniells Walk SO41 19 E4
Daniels Walk SO40 26 B3
Dark La, New Milton BH25 22 A3
Dark La, Southampton SO45 10 A4
Dark La, Christchurch BH23 14 D1
Davis Field BH25 23 A5
Davis Fld BH25 22 A4
Dawkins Way BH25 23 B5
Dayrell Cl SO40 26 B2
De Havilland Way BH23 13 F5
De La Warr Rd SO41 21 C3
De Mowbray Way SO41 19 E5
Deans Ct SO41 21 E2
Dearing Cl SO43 20 C4
Deer Park Cl BH25 22 A3
Deerhurst Cl SO40 26 C5
Deerleap Cl SO45 17 F4
Deerleap La SO40 4 C3
Deerleap Way, New Milton BH25 22 C1
Deerleap Way, Southampton SO45 17 F4
Delft Mews*, Scotts Hills La BH23 12 A1
Delta Cl BH23 13 G3
Denbigh Cl SO40 26 C6
Dene Cl BH24 25 E1
Dene Rd SO40 4 B3
Dene Way SO40 4 B3
Deneside Copse SO41 18 C5
Denham Dr BH23 14 C4
Denholm Cl BH24 25 E1
Denmead BH25 22 E3
Dennistoun Av BH23 13 E3
Denny Cl SO45 10 C2
Deridene Ct SO40 26 C5
Derritt La BH23 6 A2
Derrybrian Gdns BH25 23 C5
Derwent Dr SO40 26 B3
Derwent Rd BH25 22 C2
Deverel Cl BH23 12 A2

Devonshire Gdns SO45 17 F6
Deweys La BH24 25 A3
Diamond Cl SP6 11 E4
Diamond Ct SP6 11 E4
Diana Ct BH23 14 D5
Dibden Lodge Cl SO45 17 E2
Dickens Dell SO40 26 A5
Dilly La BH25 23 B7
Dinham Ct BH25 22 E3
Dinham Rd BH25 22 E3
Disraeli Rd BH23 12 D4
Ditchbury SO41 19 E1
Dock Cl SO41 18 C3
Dock La SO42 5 D3
Doe Copse Way BH25 22 A3
Dolphin Pl BH25 23 B8
Dominy Cl SO45 17 G2
Donnington Dr BH23 13 H3
Dorland Gdns SO40 26 C5
Dorset Rd BH23 13 E2
Douglas Way SO45 17 E2
Doveys Cl BH24 8 C3
Downs Park Av SO40 27 F6
Downs Park Cres SO40 27 F6
Downs Park Rd SO40 27 F6
Downton La SO41 21 A2
Downwood Cl, Fordingbridge SP6 11 C3
Downwood Cl, Southampton SO45 16 C5
Drake Cl, New Milton BH25 22 A4
Drake Cl, Ringwood BH24 25 E1
Drake Cl, Christchurch BH23 13 F4
Drakes Cl SO45 17 E4
Draper Rd BH23 13 E3
Drayton Pl SO40 26 D5
Driftwood Gdns SO40 26 B5
Druitt Rd BH23 13 E2
Drummond Rd SO45 17 F2
Dryden Pl SO41 21 D2
Duart Ct BH25 22 D4
Duck Island La BH24 25 B4
Ducking Stool La BH23 12 B4
Dudley Av, Lymington SO41 24 A3
Dudley Av, Fordingbridge SP6 11 E2
Dudley Pl*, Station Rd BH25 23 C5
Dukeswood Dr SO45 17 F5
Dunbar Cres BH23 14 D3
Duncan Rd BH25 22 E3
Duncliff Rd BH6 12 A6
Dunedin Gro BH23 14 A5
Dunfield Copse SO45 10 B3
Dunford Cl BH25 15 H4
Dunlin Cl BH23 13 G5
Durland Cl BH25 23 B5
Durley Cres SO40 26 D6
Durlston Rd SO16 27 H2
Durnstown SO41 6 C5
Durrant Way SO41 6 B5

Earley Ct SO41 19 F3
Earlsdon Way BH23 14 C4
East Av BH25 15 G5
East Bank Rd SO42 7 C4
East Cl BH25 15 H4
East Cliff Way BH23 14 A5
East Hill SO41 19 E3
East La SO41 24 B5
East View Rd BH24 25 C3
Eastern Rd SO41 19 E3
Eastern Way SO41 21 F3
Eastfield Ct BH24 25 D3
Eastfield La, Ringwood BH24 25 D2
Eastfield La, Ringwood BH24 25 D3
Eastlands BH25 23 D5
Eastmeare Ct SO40 26 B5
Ebenezer La BH24 25 B4
Eddystone Rd SO40 26 C1
Edmund Cl BH23 12 D4
Edmunds Cl BH25 23 B6
Edward Cl SO45 10 A4
Edward Rd, Christchurch BH23 13 F2
Edward Rd, Southampton SO45 17 F2
Edwards Cl SO41 18 C4
Edwina Cl BH24 25 D1
Efford Ct SO41 18 C5
Efford Way SO41 18 B5
Elcombes Cl SO43 20 B4
Elderberry La BH23 13 F4
Eldon Av BH25 23 A6
Eldon Cl BH25 15 H4
Elgar Ct SO41 19 F3
Elgin Cl SO45 17 G4
Eling Hill SO40 27 G6
Eling La SO40 27 F5
Elizabeth Cres SO41 24 C3
Elizabeth Gdns, Christchurch BH23 14 A5
Elizabeth Gdns, Southampton SO45 17 F5
Elkhams Cl SO41 24 A5
Elldene Ct SO40 26 D6
Ellery Gro SO41 19 E2
Ellingham Rd BH25 15 G5
Elliot Cl SO40 26 C5
Elm Av, New Milton BH25 23 B5
Elm Av, Lymington SO41 18 D5
Elm Vw BH24 25 E4
Elmers Way BH23 6 C1
Elms Cl SP6 11 A2
Elmtree Cl SO40 4 B3
Elmwood Av SP6 11 C3
Elmwood Way BH23 14 D5
Elphinstone Rd BH23 15 E4
Elvin Cl SO41 24 A1
Embankment Way BH24 25 C4
Embley Cl SO40 26 C1
Empress Rd SO43 20 C3
Emsworth Rd SO41 19 E3
Endeavour Way SO45 17 F1
Enderwood Cl SO40 26 B3
Esdaile La BH24 8 C3
Ethelred Gdns SO40 26 C5
Euston Gro BH24 25 C4
Everest Rd BH23 12 D2
Evergreens SO40 27 F5
Everlea Cl SO41 24 A4
Everon Gdns BH25 23 C5
Everton Rd, Hordle SO41 24 A1
Everton Rd, Everton SO41 24 A4
Ewart Ct SO45 17 F2
Ewell Way SO40 26 C2
Exeter Ct BH23 15 E5
Exmoor Cl SO40 26 B4
Eyeworth Walk SO45 16 B3
Eynon Mews BH24 25 B4
Eyre Cl SO40 26 D5

Fairfield BH23 12 A3
Fairfield Cl, Christchurch BH23 12 B3
Fairfield Cl, Lymington SO41 19 E4
Fairfield Cl, Southampton SO45 17 F3
Fairfield Rd BH25 15 H5
Fairlea Rd SO41 19 F2
Fairlie BH24 25 D1
Fairlie Pk BH24 25 D1
Fairmead Way SO40 4 D1
Fairmile Rd BH23 12 A2
Fairview Cl SO45 17 F4
Fairview Dr SO45 17 F3
Fairview Par SO45 17 F4
Fairway Rd SO45 17 E3
Falcon Dr BH23 13 G5
Falcon Flds SO45 10 C2
Falstaff Way SO40 26 D6
Farm Cl BH24 25 B2
Farm La BH23 13 G5
Farm La North BH25 23 B7
Farm La South BH25 23 B7
Farmdene Cl BH23 14 B4
Farmers Walk SO41 24 A5
Farnleys Mead SO41 19 E5
Fathersfield SO42 7 C3
Fawcett Rd BH25 23 A5
Fawley By-Pass SO45 10 C2
Fawley Rd, Fawley SO45 10 A2
Fawley Rd, Hythe SO45 17 F6
Fawn Gdns BH25 22 A3
Felton Cres BH23 14 D4
Fennel Gdns SO41 19 E1
Fern Rd SO45 17 E4
Ferndale Rd BH25 22 C2
Fernglade BH25 22 C3
Fernhill La BH25 22 B1
Fernhill Rd BH25 22 B3
Fernhills Rd SO45 17 G5
Fernlea Way SO45 16 C4
Fernleigh Cl BH25 23 C6
Ferry Rd SO45 17 E3
Fibbards Rd SO42 7 C3
Field Pl BH25 15 G4
Field Walk SO41 18 D2
Field Way BH23 14 B4
Fieldfare Ct SO40 26 B3
Fieldway BH24 25 C2
Fifth St SO45 10 A2
Filton Cl SO40 26 B3
Filton Rd SO41 18 D2
Fir Av BH25 23 C5
Fir Cl SO43 20 B4
Fir Rd SO40 4 A3
Fir Tree La BH23 14 B3
Fir Tree Rd SO40 9 C2
Fire Station La SO42 5 C4
Firmount Cl SO41 24 B5
Firshill BH23 14 C4
First Marine Av BH25 23 B8
First St SO45 10 C2
Firtree Cres SO41 24 A2
Fishermans Bank BH23 13 E5
Fishermans Quay SO41 19 F3
Fishers Rd SO40 27 F5
Flambard Av BH23 12 A1
Flaxfields End SP6 11 D4
Fletcher Cl SO45 16 C3
Fletchwood La SO40 4 A3
Fletchwood Rd SO40 26 B5
Fleuret Cl SO45 17 F6
Floriston Gdns BH25 22 E4
Flowerdown Cl SO40 26 B3
Flushards SO41 19 G4
Foldsgate Cl SO43 20 B2
Fontwell Cl SO40 26 C2
Fordingbridge By-Pass SP6 11 F4
Forest Cl, Brockenhurst SO42 7 B3
Forest Cl, Christchurch BH23 14 B3
Forest Ct BH25 22 C4
Forest Edge SO45 10 C2
Forest Front SO45 17 F6
Forest Gate SO45 10 B6
Forest Gate Gdns SO41 19 E6
Forest Gdns SO43 20 B3
Forest Glade Cl SO42 7 A3
Forest Hall SO42 7 D3
Forest Hill Way SO45 17 E5
Forest Hills Ct BH24 25 F4
Forest Oak Dr BH25 22 C2
Forest Park Rd SO42 7 B2
Forest Pines BH25 22 C2
Forest Rd BH24 8 A1
Forest Rise BH23 14 B3
Forest Vw SO42 7 A3
Forest Way, Christchurch BH23 14 A3
Forest Way, Lymington SO41 24 A4
Foresters Rd SO45 10 A3
Forestlake Av BH24 25 F4
Forestside Gdns BH24 25 E1
Forge Rd SO45 10 A6
Forward Dr SO41 18 D5
Fountain Way BH23 12 B4
Fourshells Cl SO45 10 A3
Fourth St SO45 10 A2
Fox Fld SO41 24 B5
Fox Lease Ter SO43 20 B5
Fox Pond La SO41 18 D5
Foxbury Cl SO45 17 F5
Foxcote Gdns BH25 22 A4
Foxglove Cl BH23 13 H2
Foxglove Pl BH25 22 E3
Foxhayes La SO45 10 A6
Foxhills SO40 4 C2
Foxhills Cl SO40 4 B2
Foxlands SO45 10 A6
Foxs Walk SO45 10 A6
Foxtail Dr SO45 17 E5
Foxwood Av BH23 13 E5
Foxy Paddock SO45 10 A6
Frampton Cl BH25 22 D2
Frampton Pl BH24 25 B3
Frampton Way SO40 27 E5
Frances Ct BH23 15 E5
Franklin Rd BH25 22 D3
Fraser Ct*, Marryat Rd BH25 22 A4
Frayslea SO45 17 F5
Freshwater Rd BH23 14 A6
Friars Cft SO40 26 B1
Friars Rd BH23 13 H4
Friars Walk BH25 23 B6
Fritham Cl SO40 26 C4
Frobisher Cl, Ringwood BH24 25 E2
Frobisher Cl, Christchurch BH23 13 F4
Froghall SO45 17 F5
Fromond Cl*, Campion Way SO41 19 E1
Frost La SO45 17 G4
Fry Cl SO45 10 B3
Frys La SO41 24 A4
Fullerton Rd SO41 18 D3
Fulmar Dr SO45 17 G4
Fulmar Rd BH23 13 G5
Furze Cft BH25 23 B5
Furzedale Gdns SO45 17 G5
Furzedale Pk SO45 17 G5
Furzedown Mews SO45 17 G5
Furzey Av SO45 17 G4
Furzey Cl SO45 10 A4
Furzy Whistlers Cl BH23 6 C1

Gainsborough Av BH25 22 C2
Gainsborough Ct SO41 18 D5
Galsworthy Rd SO40 26 C5
Gannet Cl SO45 17 G4
Garden City SO45 16 A1
Garden Cl, Lynhurst SO43 20 A3
Garden Cl, New Milton BH25 23 B5
Garden Rd BH24 8 C3
Gardner Rd BH24 25 D4
Garendon Ct SP6 11 E2
Garrow Dr SO41 19 E2
Genoa Cl SO41 18 C6
George Rd SO41 21 D2
George St SO45 10 A5
Georgian Cl BH24 25 C2
Germaine Cl BH23 14 D4
Gilbert Cl SO41 19 E5
Gillingham Rd SO41 21 E4
Gilpin Hill SO41 6 B5
Gilpin Pl SO41 6 B4
Gipsy La BH24 25 D2
Gladstone Cl BH23 12 D4
Glebefields SO41 21 D3
Glen Cl BH25 15 G4
Glen Dr BH25 15 F5
Glen Rd SO40 27 F4
Glenavon BH25 23 D5
Glenavon Rd BH23 14 D3
Glendale Rd BH6 12 A6
Glendales BH25 15 G4
Glengarry BH25 23 D5
Glengarry Way BH23 13 H4
Glenside, Southampton SO45 17 E4
Glenside, New Milton BH25 15 F5
Glenville Cl BH23 15 E2
Glenville Rd BH23 15 E3
Glyn Jones Cl SO45 10 B3
Gold Mead Cl SO41 19 F4
Goldcrest La SO40 26 B4
Golden Cres SO41 24 A5
Golden Hind Pk SO45 17 E4
Goldfinch Cl BH25 23 A5
Goldsmith Cl SO40 26 C5
Goodwood Gdns SO40 26 C3
Goose Grn SO43 20 C5
Gooseberry La BH24 25 B3
Gordon Mount BH23 15 E4
Gordon Rd, Lymington SO41 18 C4
Gordon Rd, Christchurch BH23 15 E4
Gordon Way BH23 12 C1
Gore Grange BH25 23 A5
Gore Rd BH25 15 G2
Gorley Rd BH24 25 D2
Gorse Cl BH25 22 E3
Gorsefield Rd BH25 22 D2
Gosport La SO43 20 C5
Gosport St SO41 19 F2
Gover Rd SO16 2
Graddidge Way SO40 2
Grafton Cl BH23 1
Grafton Gdns SO41 1
Grange Cl SO41 2
Grange Rd BH23 1
Grasmere Gdns BH25 2
Gravel La, Ringwood BH24 2
Gravel La, Ringwood BH24 2
Grays Av SO45 17
Great Mead SO43 2
Greatwood Cl SO45 1
Grebe Cl, Lymington SO41 2
Grebe Cl, Christchurch BH23 1
Green Acre BH25 2
Green Acres BH23 1
Green Cl SO45 1
Green La, Cadnam SO40
Green La, Fawley SO45 1
Green La, Ringwood BH24 2
Green La, New Milton BH25 2
Green La, Fordingbridge SP6 1
Green La, Fordingbridge SP6 1
Greenacres SO45 1
Greenbanks Cl SO41 2
Greenfield Gdns BH25 2
Greenfields Av SO40 2
Greenfields Cl SO40 2
Greenfinch Walk BH24 2
Greenmead Av SO41 2
Greenside Ct BH25 2
Greenway Cl SO41 1
Greenways, Lymington SO41 2
Greenways, Christchurch BH23 1
Greenways Rd SO42
Greenwoods BH25 2
Gregory Gdns SO40 2
Grenville Cl BH24 2
Grenville Gdns SO45 1
Grigg La SO42
Grosvenor Mews SO41 1
Grove Pastures SO41 1
Grove Pl SO41 1
Grove Rd, New Milton BH25 2
Grove Rd, Lymington SO41 1
Groveley Rd BH23 1
Guillemot Cl SO45 17
Guys Cl BH24 2

Haarlem Mews BH23 1
Haglane Copse SO41 1
Haking Rd BH23 1
Hale Av BH25 2
Hale Gdns BH25 2
Halifax Way BH23 1
Halton Cl BH23
Haltons Cl SO40 2
Hambert Way SO40 2
Hamilton Cl BH23 1
Hamilton Ct SO41 2
Hamilton Mews SO45 17
Hamilton Way BH25 2
Hammonds Cl SO40 2
Hammonds Grn SO40 2
Hammonds La SO40 2
Hammonds Way SO40 2
Hampton Cl SO45 1
Hampton Dr BH24 2
Hampton La SO45 1
Hamtun Cres SO40 2
Hamtun Gdns SO40 2
Handley Ct BH24 2
Harbour Cres BH23 1
Harbour Ct BH25 1
Harbour View Ct*, Princess Av BH23 1
Hardy Cl, New Milton BH25 2
Hardy Cl, Christchurch BH23 1
Hardy Dr SO45 17

are La, Lymington SO41 22 F3
are La, New Milton BH25 22 F4
arford Cl SO41 18 C6
arold Cl SO40 26 C5
arriers Cl BH23 14 B4
arrington Ct BH23 14 D5
arris Way BH25 22 D1
arry Barrows Cl BH24 25 B3
art Cl BH25 22 A3
art Hill SO45 17 H5
artley Cl SO45 17 F6
artley Walk SO45 17 F5
arts Way SO41 24 A4
artsgrove Av SO45 10 A5
artsgrove Cl SO45 10 A4
arvester Way SO41 19 E1
arvey Gdns SO45 17 G3
arwood Cl SO40 26 D4
arwood Ct BH25 22 A4
aselbury Rd SO40 27 E4
askells Cl SO43 20 B4
aslemere Av BH23 14 D4
aslemere Pl BH23 14 D4
atchet La SO42 5 A5
atfield Ct BH25 22 A3
avelock Way BH23 14 B2
aven Ct SO41 21 C3
aven Gdns BH25 23 C5
awkhill SO45 16 B3
awkins Cl, Ringwood BH24 25 E2
awkins Cl, Christchurch BH23 13 F5
aworth Cl BH23 12 A1
awthorn Cl BH25 22 D3
awthorn Dr SO41 6 B5
awthorn Rd, Christchurch BH23 13 E2
awthorn Rd, Southampton SO45 17 E3
awthorne Rd SO40 26 C3
aydens Ct SO41 19 F3
aydock Cl SO40 26 C3
ayley Cl SO45 17 F6
aynes Way SO45 16 D5
aysoms Cl BH25 23 C6
ayward Cl SO40 26 C4
azel Cl BH23 14 A3
zel Ct*, Station Rd BH25 23 C5
zel Farm Rd SO40 26 B4
zel Rd SO41 18 B3
zelwood Av BH25 22 A3
art of Oak Mews SO41 19 E3
ath Rd, Lymington SO41 24 A2
ath Rd, Christchurch BH23 15 E3
ather Cl, Lymington SO41 24 B2
ather Cl, Southampton SO40 26 D5
ather Cl, Christchurch BH23 15 E3
ather Rd SO45 10 A3
atherstone Av SO45 17 E6
athfield SO45 17 E4
athwood Av BH25 15 H4
athy Cl BH25 23 A7
dgerley BH25 23 C7
dley Cl SO45 10 A4
mming Cl SO40 27 E5
ngistbury Rd BH25 15 H4
rbert Rd BH25 22 C4
ron Cl SO41 6 B5
rons Wood SO40 26 C2
tford Cl SP6 11 E2
tford Ct BH23 15 E5
es Hill La SO42 5 B1
h Ridge Cres BH25 22 D4
h St, Ringwood BH24 25 A3
h St, Brockenhurst SO42 5 B4
h St, Christchurch BH23 12 B4
h St, Lynhurst SO43 20 B4
h St, Lymington SO41 19 E3
h St, Milford on Sea SO41 21 E3

High St, Fordingbridge SP6 11 E4
High St, Hythe SO45 17 F2
High St, Southampton SO40 27 F5
Highbank Gdns SO45 11 E4
Highbury Cl BH25 22 C4
Highcliffe Rd BH23 13 G2
Highfield SO41 18 D4
Highfield Av, Ringwood BH24 25 B2
Highfield Av, Lymington SO41 18 D4
Highfield Cl SO41 6 B5
Highfield Dr BH24 25 C1
Highfield Gdns SO41 6 B5
Highfield Rd, Ringwood BH24 25 C2
Highfield Rd, Lymington SO41 18 C3
Highfold Ct SO45 17 E2
Highgrove Cl SO40 4 C1
Highland Av BH23 15 E4
Highlands Cl SO45 17 F5
Highlands Rd BH25 23 B7
Highlands Way SO45 17 F5
Hightown Gdns BH24 25 D4
Hightown Hill BH24 25 F4
Hightown Rd BH24 25 B4
Highwood Rd SO42 7 C4
Hill Cl BH23 6 B3
Hill La BH23 6 B3
Hillary Cl SO43 20 C6
Hillary Rd BH23 13 E2
Hillditch SO41 19 E1
Hillside Rd SO41 18 D4
Hillview Rd SO45 17 E3
Hiltom Rd BH24 25 C3
Hilton Rd BH25 22 D3
Hinton Wood Av BH23 14 C2
Hinton Wood La BH23 14 B2
Hirst Rd SO45 17 G4
Hives Way SO41 19 E1
Hobart Dr SO45 17 G3
Hobart Rd BH25 23 B5
Hoburne Gdns BH23 14 A4
Hoburne La BH23 13 H2
Holes Cl SO41 24 A1
Holland Rd SO40 26 B5
Hollands Wood Dr BH25 22 C2
Hollies Cl SO41 6 B6
Holly Gdns, Lymington SO41 21 C3
Holly Gdns, Christchurch BH23 12 D1
Holly Hatch Rd SO40 26 D5
Holly La, New Milton BH25 22 E3
Holly La, Christchurch BH23 15 F2
Holly Rd, Ashurst SO40 4 A4
Holly Rd, Fawley SO45 10 A5
Hollybank Cl SO45 17 F3
Hollybank Cres SO45 17 F2
Hollybank Rd SO45 17 E3
Hollywood La SO41 19 E2
Holm Cl BH24 25 E1
Holme Cl SO41 18 C5
Holme Rd BH23 15 E4
Holmhurst Av BH23 14 C3
Holmsley Cl SO41 18 B5
Holmsley Ct SO40 26 B3
Holmwood Garth BH24 25 E4
Home Farm Cl SO45 17 G4
Homefield SO43 20 C3
Homegrange Ho SO41 21 D4
Homewood Cl BH25 22 D4
Honey La BH24 8 B2
Honeybourne Cres BH6 12 A6
Honeysuckle Gdns SO41 24 A5
Honeysuckle Way BH23 13 G2
Honeywood Cl SO40 26 C2
Hordle La SO41 24 B3
Horlock Rd SO42 7 C3
Hotspur Cl SO45 17 E2
Hounsdown Av SO40 27 E6
Hounsdown Cl SO40 27 E6
House Mews*, Ellery Gro SO41 19 E2
Howard Cl, Ringwood BH24 8 C3

Howard Cl, Christchurch BH23 13 F5
Howards Mead SO41 18 C4
Howe Cl, New Milton BH25 22 A4
Howe Cl, Christchurch BH23 13 F5
Howlett Cl SO41 18 D2
Hudson Cl BH24 25 E2
Hudson Ct SO40 26 C5
Hungerfield Cl BH23 6 B2
Hunt Rd BH23 13 E2
Hunter Cl BH23 13 G3
Hunters Cres SO40 26 B5
Hunters Hill SO40 4 D2
Huntingdon Cl SO40 26 D2
Hurdles Mead SO41 21 E4
Hursley Dr SO45 10 A6
Hurst Cl, Christchurch BH23 15 F2
Hurst Cl, Southampton SO40 27 F3
Hurst Ct, Lymington SO41 21 C3
Hurst Ct, Christchurch BH23 15 E5
Hurst Rd, Ringwood BH24 25 B1
Hurst Rd, Lymington SO41 21 E4
Hurstbourne Av BH23 14 C3
Huxley Ct SO45 16 C4
Hyde Cl, Southampton SO40 26 B4
Hyde Cl, Lymington SO41 6 C5
Hyman Way SO40 27 E4
Hynesbury Rd BH23 13 H4
Hythe By-Pass SO45 16 A1
Hythe Marina SO45 17 F1

Ibbotson Way SO40 4 D1
Imber Dr BH23 14 D4

INDUSTRIAL & RETAIL:
Airfield Ind Est BH23 13 F4
Ambassador Ind Est BH23 13 F4
Ashford Works Ind Est SP6 11 C3
Avon Trading Pk BH23 12 A3
Beaver Ind Est BH23 13 F4
Carvers Trading Est BH24 25 B3
Crow Arch La Ind Est BH24 25 C4
Gore Rd Ind Est BH25 22 A4
Grange Rd Bsns Centre BH23 13 G3
Griffin Ind Pk SO40 26 D1
Hightown Ind Est BH24 25 C4
Hughes Bsns Centre BH23 13 F3
Lulworth Bsns Pk SO40 26 D1
Milford Trading Est SO41 21 E3
New Forest Enterprise Centre SO40 26 D6
Priory Ind Pk BH23 13 G3
Ringwood Trading Est BH24 25 C4
Rushington Bsns Pk SO40 26 D6
Saxon Centre BH23 12 B4
Saxon Sq BH23 12 B4
Sea Vixen Ind Est BH23 13 F3
Silver Bsns Pk BH23 13 E3
Somerford Bsns Pk BH23 13 F3
South Hampshire Ind Pk SO40 26 D1
Sway Pk Ind Est SO41 6 B6
Test Valley Bsns Centre SO16 27 G1
The Shipyard Ind Est SO45 17 G2
Wick 1 Ind Est BH25 22 A4
Wick 2 Ind Est BH25 15 H2
Williams Ind Pk BH25 15 H2
Yeoman Ind Pk SO16 27 H2

Ingle Glen SO45 17 F5
Ingle Grn SO40 26 B3
Inglegreen Cl BH25 23 A6
Inglewood Dr BH25 23 D5
Inveravon BH23 13 F6
Ipley Way SO45 17 F5
Irvine Way BH23 13 E2
Island View Av BH23 13 G4
Island View Cl SO41 21 E4
Island View Ct BH25 23 B7
Island View Rd BH25 15 G5
Itchin Cl SO40 26 D5
Ivy Cl SO40 26 D2

Jackie Wigg Gdns SO40 27 F5
Jacmar Ct*, Ashley Rd BH25 22 C4
Jacobean Cl BH23 15 E3
Janes Cl SO45 10 A5
Janred Ct BH25 15 H5
Jasmine Ct SO41 19 E3
Jaundrells Cl BH25 22 D4
Jays Ct BH23 15 E5
Jellicoe Dr BH23 13 F4
Jennings Rd SO40 27 F4
Jesmond Av BH23 14 D5
Jessop Cl SO45 17 F1
Jessop Walk SO45 17 E1
Johnstone Rd BH23 13 E4
Jonathan Cl SO41 19 E2
Jones La SO45 17 F2
Jordans La SO41 6 C4
Jowitt Dr BH25 23 A5
Joyce Dickson Cl BH24 25 D4
Jubilee Cl, Fordingbridge SP6 11 C3
Jubilee Cl, Ringwood BH24 25 D2
Jubilee Ct SO41 6 B6
Jubilee Rd SP6 11 B3
Julia Cl BH23 14 C5
Junction Rd SO40 27 F4
Juniper Cl SO41 18 C5

Kay Cl BH23 13 E4
Kayleigh Cl SO40 26 C5
Keats Av SO41 21 D3
Kelsall Gdns BH25 22 B4
Kelvin Cl SO45 17 F3
Kenilworth Cl BH25 22 C4
Kenilworth Ct*, Stour Rd BH23 12 A3
Kenmore Cl SO40 4 D1
Kennard Ct BH25 22 B4
Kennard Rd BH25 22 A3
Kenneth Ct BH23 15 E5
Kennington La SO40 9 D2
Kensington Flds SO45 16 D4
Kensington Pk SO41 21 D3
Kent Gdns SO40 26 D6
Keppel Cl BH24 25 D3
Kerry Cl SO41 18 C4
Kerry Gdns SP6 11 A3
Kestrel Ct BH24 25 B2
Kestrel Dr BH23 13 G4
Keswick Ct BH25 22 C2
Keswick Rd BH25 22 C2
Kewlake La SO40 9 A1
Keyes Cl BH23 13 F4
Keyhaven Rd SO41 21 F3
Keysworth Av BH25 23 A7
Kilmington Way BH23 14 C4
Kiln Cl SO45 17 E4
Kilnyard Cl SO40 26 C3
King George Mobile Home Pk BH25 23 A6
Kingfisher Way, Ringwood BH24 25 D1
Kingfisher Way, Christchurch BH23 13 F5
Kings Arms La BH24 25 A3
Kings Arms Row BH24 25 A3
Kings Av BH23 12 A4
Kings Cl, Lynhurst SO43 20 B3
Kings Cl, Lymington SO41 18 D3
Kings Cres SO41 18 D2
Kings Farm La SO41 24 C3
Kings Rd, Lymington SO41 18 D3
Kings Rd, New Milton BH25 22 D3
Kings Ride SO45 10 A6
Kings Saltern Rd SO41 19 G5
Kingsbere Gdns BH23 14 D4
Kingsbury La BH24 25 A3

Kingsfield, Ringwood BH24 25 C3
Kingsfield, Lymington SO41 19 G5
Kingsley Cl BH6 12 A6
Kingsley Gdns SO40 26 B5
Kingston Pk SO41 18 D5
Kinross Rd SO40 27 E5
Kitchers Cl SO41 6 B4
Kitwalls La SO41 21 D2
Kivernell Pl*, Kivernell Rd SO41 21 C3
Kivernell Rd SO41 21 C3
Knapp Cl BH23 12 A2
Knapp Mill Av BH23 12 A2
Knellers La SO40 4 C1
Knightcrest Pk SO41 24 C5
Knighton Pk BH25 15 H4
Knightstone Grange SO45 17 F4
Knightwood Av SO43 20 B3
Knightwood Cl, Lynhurst SO43 20 B3
Knightwood Cl, Southampton SO40 4 B3
Knightwood Cl, Christchurch BH23 14 B4
Knightwood Rd SO45 17 G4
Knowland Dr SO41 21 E2
Knowle Rd SO42 7 B3
Knowles Cl BH23 12 D3

Laburnum Dr SO41 24 B5
Lackford Av SO40 27 E5
Ladycross Rd SO45 17 F5
Ladysmith Cl BH23 12 D3
Lake Grove Rd BH25 22 B3
Lakeside BH24 25 E4
Lakeview Dr BH24 25 E4
Lakewood Rd, Southampton SO40 4 B2
Lakewood Rd, Christchurch BH23 14 C3
Lambourne Cl SO45 17 E5
Lammas Rd SO45 17 G5
Lancaster Cl BH23 13 H3
Lanehays Rd SO45 17 E3
Langdown Firs SO45 17 F4
Langdown Lawn SO45 17 F4
Langdown Lawn Cl SO45 17 F4
Langdown Rd SO45 17 G3
Langley Gdn SP6 11 F2
Langley Lodge Gdns SO45 10 A6
Langley Rd BH23 14 B3
Langton Cl BH25 23 C6
Lapwing Dr SO40 26 B3
Larch Cl SO41 24 A1
Larchwood Rd SO40 26 B5
Lark Rd BH23 13 G5
Larkshill Cl BH25 22 C3
Latch Farm Av BH23 12 A2
Latchmore Dr SO45 16 C3
Latimers Cl BH23 14 C3
Laundry La SO41 21 F3
Laurel Cl, Lymington SO41 24 A1
Laurel Cl, Christchurch BH23 14 A3
Laurel Cl, Southampton SO45 17 E3
Lavender Rd SO41 24 A2
Lawford Way SO40 26 D4
Lawn Cl SO41 21 F3
Lawn Rd, Lymington SO41 18 C4
Lawn Rd, Milford on Sea SO41 21 F3
Lawrence Rd BH24 25 E1
Le Patourel Cl BH23 12 D3
Lea Cl SO43 20 C4
Lebanon Rd SO15 27 H4
Ledbury Rd BH23 13 E5
Leelands SO41 18 D5
Leigh Pk SO41 18 D3
Leigh Rd BH25 22 B3
Lentune Way SO41 19 E5
Lepe Rd SO45 10 A6
Lewis Cl SO45 16 C4
Lexby Rd SO40 27 F6
Leyside BH23 13 F3
Library Rd SO40 27 F4
Lightning Cl SO45 10 B3
Lilac Cl BH24 25 C2
Lime Cl SO45 16 D5
Lime Gro SO41 24 B5

Lime Walk SO45 16 D5
Linbrook Ct BH24 25 B1
Linda Rd SO45 10 C2
Linden Ct*, Linden Gdns BH24 25 A2
Linden Gdns BH24 25 B2
Linden Way SO41 18 C3
Lineside BH23 12 C2
Linford Cl BH25 22 C3
Linford Rd BH24 25 E1
Lingwood Av BH23 13 E4
Linhorns La BH25 22 C1
Link Rd BH24 25 D2
Linnet Cl BH24 25 E4
Linnet Ct BH25 22 A4
Linwood Cl SO45 17 F5
Lisle Cl SO41 19 E4
Lisle Court Rd SO41 19 H3
Litchford Rd BH25 22 D3
Little Barrs Dr BH25 22 C3
Little Burn SO41 6 C4
Little Dene Copse SO41 18 C5
Little Reynolds SO40 26 C6
Livingstone Rd BH23 12 D3
Lockerley Cl SO41 19 E5
Locks Cotts SO45 16 C1
Lodge Dr SO45 17 F5
Lodge Rd SO41 18 C4
Long Beech Dr SO40 26 D5
Longbridge Cl SO40 26 C1
Longbridge Ct SO40 26 C1
Longfield Rd SO41 24 C3
Longford Pl SO41 18 D6
Longleat Gdns BH25 15 H1
Longmead Rd BH24 8 B2
Longmeadow Gdns SO45 17 F3
Longstock Cres SO40 26 D4
Loperwood SO40 26 A1
Loperwood La, Southampton SO40 26 A1
Loperwood La, Southampton SO40 26 A2
Loraine Av BH23 15 F4
Lovage Gdns SO40 26 B4
Love La SO41 21 E3
Lower Ashley Rd BH25 22 E4
Lower Bartons SP6 11 E3
Lower Brownhill Rd SO16 27 H2
Lower Buckland Rd SO41 18 D1
Lower Mead End Rd SO41 6 A6
Lower Mullins La SO45 17 E3
Lower Pennington La SO41 18 D6
Lucerne Rd SO41 21 E3
Lumby Dr BH24 25 D2
Lumby Dr Caravan Pk BH24 25 D2
Lunedale Rd SO45 16 D6
Lychgate Ct BH24 25 E4
Lydgate SO40 26 C5
Lydlynch Rd SO40 27 E4
Lyme Cres BH23 14 D5
Lymefields SO41 21 E2
Lymington Rd, Christchurch BH23 14 B5
Lymington Rd, Everton SO41 24 B5
Lymington Rd, New Milton BH25 23 B6
Lymington Rd, Brockenhurst SO42 7 D6
Lymington Rd, Milford on Sea SO41 21 E2
Lymore La, Everton SO41 24 B6
Lymore La, Milford on Sea SO41 21 F1
Lymore Valley SO41 21 F1
Lyndale Cl SO41 21 E3
Lyndhurst Rd, Beaulieu SO42 5 A1
Lyndhurst Rd, Ashurst SO40 4 A5
Lyndhurst Rd, Cadnam SO40 9 B2
Lyndhurst Rd, Brockenhurst SO42 7 D1
Lyndhurst Rd, Ringwood BH24 8 E1
Lyndhurst Rd, Christchurch BH23 13 G2
Lynes La BH24 25 A3
Lynric Cl BH25 23 B8
Lynton Ct SO40 26 D6
Lyon Av BH25 22 C4
Lysander Cl BH23 13 H3
Lyster Rd SP6 11 F3
Lyteltane Rd SO41 18 D5
Lytton Rd SO45 17 G4

Madeira Walk SO41 19 F3
Magdalen La BH23 12 A4
Magnolia Cl, Bournemouth BH6 12 A6
Magnolia Cl, Southampton SO45 16 C3
Magpie Dr SO40 26 C4
Magpie Gro BH25 23 A5
Maiden La SO41 19 F6
Main Rd, Hythe SO45 16 A1
Main Rd, Hythe SO45 16 C1
Main Rd, Southampton SO40 4 D2
Mallard Cl, Lymington SO41 24 C2
Mallard Cl, Christchurch BH23 13 G4
Mallory Cl BH23 13 F2
Mallow Cl BH23 13 H2
Malvern Ct*, Dorset Rd BH23 13 F2
Malvern Dr SO45 16 D4
Malwood Gdns SO40 26 B4
Malwood Rd SO45 17 E2
Malwood Rd West SO45 17 E3
Manchester Rd SO41 6 B4
Manderley SO41 21 E4
Manning Av BH23 14 A3
Manor Cl, Lymington SO41 21 D2
Manor Cl, Southampton SO40 26 D5
Manor Ct, Ringwood BH24 25 B2
Manor Ct, Fordingbridge SP6 11 F3
Manor Farm Cl BH25 23 A5
Manor Farm Rd SP6 11 B3
Manor Gdns BH24 25 B3
Manor Rd, Southampton SO45 16 A2
Manor Rd, Christchurch BH23 12 A4
Manor Rd, Ringwood BH24 25 C3
Manor Rd, New Milton BH25 22 C4
Manor Rd, Lymington SO41 21 D2
Mansell Cl SO45 16 D6
Mansfield Rd BH24 25 B3
Maple Cl, New Milton BH25 23 C8
Maple Cl, Christchurch BH23 14 D5
Maple Gdns SO40 26 B5
Maplewood Cl SO40 26 B5
Marabout Cl BH23 12 D3
Marbrean Cl SP6 11 C3
Marchwood By-Pass SO40 27 E6
Marine Ct BH25 15 G5
Marine Dr BH25 23 A8
Marine Dr East BH25 23 B8
Marine Dr West BH25 15 G5
Mariners Ct, Christchurch BH23 13 F4
Mariners Ct, Lymington SO41 19 F4
Mariners Mews SO45 17 G3
Market Pl, Ringwood BH24 25 A3
Market Pl, Fordingbridge SP6 11 E4
Marks La BH25 22 B1
Marl La SP6 11 B2
Marlborough Ct SO45 17 E4
Marlborough Pl SO41 18 D2
Marley Av BH25 22 A3
Marley Cl BH25 22 A4
Marlpit Dr BH23 15 E3
Marmion Grn BH23 13 E3
Marram Cl SO41 19 E1
Marryat Ct, New Milton BH25 22 A3
Marryat Ct, Christchurch BH23 15 E5
Marryat Rd BH25 22 A4
Marsh Ditch BH23 12 A4
Marsh La, Southampton SO45 10 C1
Marsh La, Christchurch BH23 12 D4
Marsh La, Lymington SO41 19 E1
Marsh Par SO45 17 F2
Marston Cl BH25 22 D2
Marston Gro BH23 14 C3
Marston Rd BH25 22 C2
Martins Hill Cl BH23 12 C1
Martins Hill La BH23 12 C1
Martins Rd SO42 7 D2
Mary Mitchell Cl*, Kings Arms La BH24 25 A3
Marybridge Cl SO40 26 D6
Maryland Ct SO41 21 B3
Maryland Gdns SO41 21 C3
Masterson Cl BH23 12 D3
Matley Gdns SO40 26 B4
Maturin Cl SO41 19 E4
May Av SO41 19 E2
May Gdns BH23 15 E3
Mayfield Av SO40 27 E4
Mayfield Rd SP6 11 B3
Mayflower Cl SO41 19 G4
Mayfly Cl SP6 11 E2
Maynard Rd SO40 27 E4
Meacher Cl SO40 26 D4
Mead End Rd SO41 6 A5
Mead Rd SO41 18 C5
Meadow Av SP6 11 E3
Meadow Cl, Christchurch BH23 6 B3
Meadow Cl, Burley BH24 8 B4
Meadow Cl, Ringwood BH24 25 C2
Meadow Cl, Fordingbridge SP6 11 E2
Meadow Cl, Southampton SO40 27 E6
Meadow Ct SP6 11 E3
Meadow End SO42 7 B3
Meadow Rd, Ringwood BH24 25 C2
Meadow Rd, New Milton BH25 22 C3
Meadow Rd, Lymington SO41 18 C5
Meadow Way, Ringwood BH24 25 C2
Meadow Way, Southampton SO45 10 C2
Meadow Way, New Milton BH25 23 C8
Meadowland BH23 13 E4
Meadowlands SO41 18 C3
Medina Way BH23 14 A6
Medlar Cl BH23 12 C1
Meerut Rd SO42 7 C3
Meeting House La BH24 25 A3
Melbury Cl SO41 18 D3
Melrose Ct, Southampton SO40 26 B3
Melrose Ct, New Milton BH25 22 E4
Mendip Cl BH25 23 D5
Mendip Ct*, Dorset Rd BH23 13 F2
Mendip Gdns SO45 16 D4
Meredith Cl BH23 12 D3
Meredith Gdns SO40 26 C5
Merley Dr BH23 14 D4
Merlin Cl BH24 25 E4
Merlin Way BH23 13 G5
Merriemeade Cl SO45 16 D6
Merriemeade Par SO45 16 D6
Merrivale Cl SO45 17 E3
Merryfield Cl BH23 6 B2
Merryweather Est BH24 25 E2
Merton Cl SP6 11 E2
Merton Ct BH23 15 E5
Merton Gro BH24 25 B2
Meyrick Cl BH23 6 B3
Michaels Way SO45 17 E2
Michigan Way SO40 26 A3
Middle Common Rd SO41 18 A3
Middle La BH24 25 B3
Middle Rd, Tiptoe SO41 6 A6
Middle Rd, Sway SO41 6 B5
Middle Rd, Lymington SO41 18 D3
Middleton Rd BH24 25 C2
Midway SO45 17 E4
Milford Cres SO41 21 E3
Milford Ct SO41 21 E3
Milford Rd, Everton SO41 24 B5
Milford Rd, Lymington SO41 18 C6
Milford Rd, New Milton BH25 23 C6
Milkwood Ct SO40 26 B4
Mill Ct SP6 11 E4
Mill La, Brockenhurst SO42 7 D4
Mill La, Ringwood BH24 8 E1
Mill La, Lymington SO41 19 F3
Mill La, Christchurch BH23 15 F4
Mill Mdw SO41 21 D3
Mill Rd, Christchurch BH23 12 A2
Mill Rd, Southampton SO40 27 F4
Mill Way SO40 26 D6
Miller Cl BH25 22 D3
Miller Rd BH23 12 D3
Millers Way SO45 17 E5
Millfield BH25 15 H1
Millhams St BH23 12 B4
Millhams St North BH23 12 B4
Milliken Cl SO45 10 B3
Millyford Cl BH25 15 G4
Milne Cl SO45 16 C4
Milton Gro BH25 23 C5
Milton Mead BH25 23 A5
Milverton Cl, Christchurch BH23 14 C3
Milverton Cl, Southampton SO40 27 F6
Milverton Rd SO40 27 F6
Minterne Rd BH23 13 E5
Mitchell Cl BH25 23 B7
Moat La BH25 23 B6
Moffat Rd BH23 12 D3
Molefields SO41 21 E3
Molyneaux Rd BH25 22 E4
Monks Pl SO40 26 D5
Monks Walk SO45 17 E6
Monkshood Cl BH23 13 H1
Monkswell Grn BH23 12 C4
Monkton La SO40 26 C6
Monmouth Gdns SO40 26 C6
Montagu Pk BH23 15 E5
Montagu Rd BH23 15 E5
Montague Ct SO45 16 D6
Monteray Dr SO41 24 B1
Montgomery Av SO40 26 D4
Monument La SO41 19 G1
Moonrakers Way BH23 14 B4
Moore Cl BH25 23 A6
Moorhill Rd BH24 8 C4
Moorland Av BH25 23 A6
Moorland Cl SO45 16 C4
Moorlands Cl SO42 7 A3
Mopley SO45 10 A6
Mopley Cl SO45 10 A5
Morant Rd BH24 25 D1
Morpeth Av SO40 27 E3
Morris Cl SO45 16 C3
Mortimer Cl, Southampton SO40 26 C2
Mortimer Cl, Christchurch BH23 13 G4
Mount Av BH25 23 C5
Mount Cl BH25 23 C6
Mount House Cl SO45 17 F1
Mount Pleasant Dr BH23 6 D1
Mount Pleasant La BH24 25 B3
Mountbatten Cl BH23 13 F5
Mountbatten Ct*, Raleigh Cl BH25 22 A4
Mountbatten Rd SO40 26 C4
Mountfield SO45 16 D2
Mousehole La SO45 17 F3
Moxhams SP6 11 E3
Mude Gdns BH23 13 F
Mudeford BH23 13 E
Mudeford Green Cl BH23 13 F
Mudeford La BH23 12 D
Mulberry Gdns SP6 11 E
Mulberry Gro SO41 24 B
Mussett Cl SO40 26 D
Myrtle Av SO40 26 C
Myrtle Cl SO41 24 A

Nada Rd BH23 14 A
Naish Est BH25 15 F/G
Naish Rd BH25 15 H
Narrow La BH24 25 E
Nash Cl SO45 16 D
Nash Rd SO45 16 C
Nea Cl BH23 14 B
Nea Rd BH23 14 E
Neacroft Cl BH25 15 G
Needles Ct SO41 21 C
Needles Point SO41 21 D
Nelson Cl BH25 22 A
Nelson Ct SO45 17 G
Nelson Dr BH23 13 F
Nelson Pl SO41 19 F
New Cotts SO16 27 H
New Ct BH24 25 A
New Forest Dr SO42 7 A
New Inn La SO40 9 D
New Inn Rd SO40 9 D
New La, New Milton BH25 22 A
New La, Lymington SO41 21 F
New Rd, Fawley SO45 10 A
New Rd, Ashurst SO40 4 E
New Rd, Southampton SO45 17
New St, Ringwood BH24 25 E
New St, Lymington SO41 19 E
New St Mews SO41 19
New Valley Rd SO41 21 D
Newbridge Rd SO40 9
Newbridge Way SO41 18 D
Newcroft Gdns BH23 12
Newenham Rd SO41 19
Newlands Cl SO45 10
Newlands Copse SO45 10
Newlands Rd, Southampton SO45 10
Newlands Rd, New Milton BH25 23
Newlands Rd, Christchurch BH23 13
Newton Rd BH25 23
Nicholas Cl BH23 15
Nicholas Rd SO45 10
Nickleby Gdns SO40 26
Nightingale Dr SO40 26
Noads Cl SO45 17
Noads Way SO45 16
Noel Cl SO42 7
Norleywood BH23 14
Norman Rd SO45 10
Normandy Cl SO41 6
Normandy Dr BH23 12
Normandy La SO41 19
Normandy Way SP6 11
Norris Gdns BH25 23
North Cl SO41 19
North Greenlands SO41 18
North Head SO41 21
North Poulner Rd BH24 25
North Rd, Brockenhurst SO42 7
North Rd, Southampton SO45 16
North St SO41 18
North Weirs SO42 7
Northampton La SO45 10
Northbourne Cl SO45 17
Northerwood Av SO43 20
Northfield Rd, Ringwood BH24 25
Northfield Rd, Lymington SO41 21
Northlands Cl SO40 26
Northlands Rd SO40 26
Northover Rd SO41 18
Norton Cl BH23 12
Nouale La BH24 25
Nursery Rd BH24 25

Jutsey Av SO40 26 D1
Jutsey Cl SO40 26 D1
Jutsey La SO40 26 D1
Jutshalling Cl SO40 26 B2
Jutwood Way SO40 26 D1

ak Cl, Lynhurst SO43 20 B4
ak Cl, Hythe SO45 16 D6
ak Cl, Totton SO15 27 H4
ak Ct*,
Pennington Cl SO41 18 C5
ak Gdns SO41 24 B5
ak La BH24 25 C2
ak Rd,
New Milton BH25 22 D4
ak Rd,
Southampton SO45 16 D6
ak Tree Par BH23 6 C2
akdene SO40 26 B4
akenbrow,
Lymington SO41 6 A4
akenbrow,
Southampton SO45 16 C4
akfield Rd,
Cadnam SO40 9 D3
akfield Rd,
Totton SO40 27 E4
aklands SO41 19 F4
aklands Av SO40 27 E4
aklands Cl SP6 11 E3
aklands Way SO45 16 C5
akleigh Cres SO40 26 D5
akleigh Way BH23 14 C5
akmount Av SO40 27 E4
akridge Rd SO15 27 H4
aktree Ct SO41 21 C3
akwood Av BH25 22 D3
akwood Ct BH25 22 C4
akwood Rd BH23 14 C3
atfield Gdns SO40 26 C2
ber Rd SO42 7 B3
berfield Rd SO42 7 A2
cknell Gro SO45 16 C3
d Barn Cl BH24 25 D3
d Brickyard Rd SP6 11 A2
d Christchurch Rd
SO41 24 A5
d Farm Walk SO41 19 E4
d Farmhouse Mews
SO41 19 E2
d Lyndhurst Rd SO40 9 B2
d Mill Ho BH24 25 A4
d Milton Grn BH25 23 A6
d Milton Rd BH25 23 B6
d Orchs SO41 19 F5
d Redbridge Rd
SO15 27 H3
d Romsey Rd SO40 9 A2
d Stacks Gdns BH24 25 D4
d Vicarage La SO41 6 C6
dbarn Cl SO40 26 B1
iver Rd SO41 18 C4
ivers Cl SO40 26 A5
chard Cl,
Christchurch BH23 12 A4
chard Cl,
Ringwood BH24 25 B2
chard Cl,
Fawley SO45 10 C2
chard Cl,
Lymington SO41 21 D2
chard Cl,
Fordingbridge SP6 11 E2
chard Cl,
Southampton SO40 27 E6
chard Gdns SP6 11 E4
chard Gro BH25 23 B6
chard Mead BH24 25 B2
chard Mews BH23 12 A4
chard Mount BH24 25 B3
chard Way SO45 17 E4
chid Way BH23 12 C3
borne Ct SO41 21 C3
borne Rd,
New Milton BH25 22 C4
borne Rd,
Southampton SO40 27 F5
prey Cl BH23 13 F5
ers Walk BH25 22 D1
erbrook SO45 17 E4
erstrand Cres SO41 21 E4
iat Cl SO40 26 B4
ey Cl BH25 23 B6
ford Ter SO41 6 C4

ddock Gdns SO41 18 D2
dget Rd BH24 25 E1

Padstow Pl SP6 11 D4
Palace La SO42 5 C4
Palmer Pl BH25 22 C3
Palmerston Av BH23 12 D4
Pamplyn Cl SO41 18 D2
Paradise La SO40 9 F4
Parham Cl BH25 22 A4
Park Av SO41 18 D2
Park Cl,
Brockenhurst SO42 7 C2
Park Cl,
Lynhurst SO43 20 C6
Park Cl,
New Milton BH25 22 D2
Park Cl,
Lymington SO41 21 E3
Park Cl,
Southampton SO45 17 G3
Park Ct SO41 21 D3
Park Gdns BH23 13 E4
Park La SO41 21 D4
Park Rd,
New Milton BH25 23 A6
Park Rd,
Lymington SO41 18 D2
Park Rd, Barton on Sea
BH25 22 D2
Park Rd,
Fordingbridge SP6 11 E3
Park Rd, Milford on Sea
SO41 21 E3
Park Vw BH25 22 B4
Parkers Cl BH24 25 E2
Parkland Dr BH25 23 A6
Parklands SO40 27 F3
Parkside,
Christchurch BH23 14 B3
Parkside,
Ringwood BH24 25 B4
Parkside,
Southampton SO40 27 E6
Parsonage Barn La
BH24 25 B3
Parsonage Cl SP6 11 E3
Parsonage Park Dr SP6 11 E2
Partridge Cl BH23 13 F5
Partridge Grn BH25 22 D1
Partridge Rd,
Brockenhurst SO42 7 C4
Partridge Rd,
Southampton SO45 17 F5
Pat Bear Cl SO15 27 H4
Pauletts La SO40 26 B1
Pauntley Rd BH23 13 E4
Peace Cl BH23 6 B2
Pealsham Gdns SP6 11 D2
Pear Tree Cl BH23 6 C2
Pearce Smith Ct BH25 23 A8
Pearmain Dr SO41 19 F5
Peartree Ct SO41 19 F5
Peartree Rd SO45 17 E4
Peckham Av BH25 22 B4
Pedlars Walk BH24 25 A3
Pegasus Av SO41 24 C3
Pelham Cl BH23 12 D4
Pelican Mead BH24 25 E4
Pemberton Rd SO43 20 C3
Pembridge Rd SP6 11 F2
Pembroke Cl SO40 27 F3
Pembroke Ct BH23 15 E5
Pendleton Gdns SO45 10 A4
Penelope Ct BH23 15 E5
Penhale Way SO40 26 D6
Penn Cl BH25 23 A6
Pennant Way BH23 13 F3
Pennine Ct*,
Hunt Rd BH23 13 F2
Pennine Gdns SO45 16 C4
Pennington Cl SO41 18 C5
Pennington Oval SO41 18 B4
Penny Hedge BH25 23 C7
Penny Way BH23 14 A6
Pennys Cl SP6 11 E2
Pennys Cres SP6 11 E2
Pennys La SP6 11 E2
Pennywell Gdns BH25 22 E3
Pentland Cl SO45 16 C5
Pentridge Way SO40 26 D6
Percy Cl SO45 17 E1
Peregrine Cl SO40 26 C5
Peregrine Rd BH23 13 F5
Perrywood Gdns SO40 26 B3
Peterscroft Av SO40 4 A4
Picket Cl SP6 11 E3
Pickwick Cl SO40 26 A5
Pikes Hill SO43 20 A2
Pikes Hill Av SO43 20 B2

Pilgrim Pk Homes
Caravan Pk BH24 25 D2
Pilgrim Pl SO41 19 E4
Pilgrims Cl BH25 22 D3
Pine Cl,
Southampton SO40 4 B2
Pine Cl, Hythe SO45 17 E5
Pine Cl,
New Milton BH25 15 H5
Pine Cres BH23 14 C5
Pinecliffe Rd BH25 15 G5
Pinehurst Av BH23 13 F5
Pinetops Cl SO41 18 B3
Pinewood Cl BH23 14 D3
Pinewood Cres SO45 17 G5
Pinewood Dr SO45 17 G5
Pinewood Rd,
Lymington SO41 24 A2
Pinewood Rd,
Christchurch BH23 14 D3
Pinkney La SO43 20 A6
Pipers Ash BH24 25 E2
Pipers Cl SO40 26 D5
Pipers Dr BH23 13 G3
Pippin Cl SO41 19 F5
Pitmore La SO41 6 D5
Pitts Pl BH25 22 F4
Plantation SO41 24 B5
Plantation Dr BH23 15 E2
Player Ct SP6 11 F2
Players Cres SO40 27 E6
Pleasance Way BH25 22 A4
Pless Rd SO41 21 B3
Plover Dr SO41 21 F4
Plover Rd SO40 26 C4
Poles La SO41 19 E6
Pollards Moor Rd
SO40 9 D1
Pond Cl BH25 22 B4
Popes La SO40 27 E4
Poplar Cl,
Bransgore BH23 6 D2
Poplar Cl,
Christchurch BH23 15 E4
Poplar Cres BH24 25 D3
Poplar La BH23 6 C2
Poplar Rd BH25 22 E3
Poplar Way BH24 25 D3
Poppy Cl BH23 13 H2
Porlock Rd SO16 27 H2
Portal Rd SO40 26 D4
Portfield Cl BH23 12 A2
Poulner Hill BH24 25 F2
Poulner Pk BH24 25 E1
Pound Cl BH24 25 B2
Pound La,
Ringwood BH24 8 A4
Pound La,
Cadnam SO40 4 D2
Pound La,
Ashurst SO40 9 E1
Pound Rd SO41 18 C4
Powell Cres SO40 27 E6
Powers Court Rd
BH25 15 H5
Powis Cl BH25 22 C4
Powlett Rd SO41 19 E4
Preston Way BH23 14 A5
Prestwood Cl BH25 23 A6
Priestlands La SO41 18 D4
Priestlands Pl SO41 19 E4
Priestlands Rd SO41 18 C4
Priestley Cl SO40 26 C5
Primrose Way BH23 13 H2
Princes Cres SO43 20 D3
Princes Pl BH25 22 D3
Princess Av BH23 12 B5
Princess Rd SO40 4 B4
Priors Cl BH23 14 A6
Priory Ct BH23 12 B5
Priory Quay BH23 12 B5
Private Rd SO41 19 F2
Promenade BH23 13 H5
Prospect Pl SO45 17 F2
Provost St SP6 11 E4
Puddleslosh La SP6 11 C2
Pullman Way BH24 25 C4
Pundle Grn SO40 9 E4
Purbeck Ct*,
Dorset Rd BH23 13 F2
Purbeck Rd BH25 15 G5
Purewell BH23 12 C4
Purewell Cl BH23 12 D4
Purewell Cross Rd
BH23 12 C3
Purewell Ct BH23 12 D3
Pylewell Rd SO45 17 F2

Pyrford Mews SO41 19 E5

Quantock Ct*,
Hunt Rd BH23 13 F2
Quarr Ho SO41 6 B4
Quay Rd BH23 12 B4
Quay St SO41 19 F3
Queen Elizabeth Av
SO41 18 D3
Queen Katherine Rd
SO41 19 F3
Queen St SO41 19 E3
Queens Av BH23 12 B5
Queens Cl SO45 17 G4
Queens Ct BH25 22 E3
Queens Gdns SP6 11 E2
Queens Gro BH25 22 E3
Queens Par SO43 20 C3
Queens Rd,
Lynhurst SO43 20 D3
Queens Rd,
Christchurch BH23 12 D4
Queens Way BH24 25 C3
Queensway BH25 15 H2
Quinton Cl BH23 14 C4
Quomp BH24 25 B3

Racecourse Vw SO43 20 C3
Racketts SO45 17 F3
Radleigh Gdns SO40 26 B3
Railway Cotts SO15 27 H4
Raleigh Cl,
New Milton BH25 22 A4
Raleigh Cl,
Ringwood BH24 25 E2
Raleigh Cl,
Christchurch BH23 13 F5
Ramley Rd SO41 18 A1
Ramsey Ct BH25 22 A4
Randall Cl SO40 26 C1
Randalls La BH24 8 A1
Ranelagh Rd BH23 14 C5
Ranfurly Gdns SO45 17 E5
Ranmore Ct SO45 16 B3
Ratcliffe Rd SO45 17 F5
Raven Way BH23 13 F5
Ravens Croft Rd SO41 18 D1
Ravens Way SO41 21 D4
Ravenscourt Rd SO41 19 E4
Redan Cl BH23 14 D5
Redbridge Causeway
SO40 27 G4
Redbridge Flyover
SO16 27 H3
Redbridge La,
Southampton SO16 27 G3
Redbridge La,
Southampton SO16 27 H2
Redbridge Rd SO15 27 H3
Redbridge Towers
SO16 27 H3
Redvers Cl SO41 19 E4
Redvers Rd BH23 12 D3
Redwing Gdns SO40 26 B3
Redwood Cl,
Ringwood BH24 25 C3
Redwood Cl,
Southampton SO45 16 C4
Redwood Cl,
Lymington SO41 18 D2
Redwood Gdns SO40 26 C5
Regency Pl BH24 25 B2
Regent Way BH23 12 B4
Reid St BH23 12 A3
Renouf Cl SO41 18 C4
Reynolds Dale SO40 4 C1
Rhinefield Cl SO42 7 B3
Rhinefield Rd SO42 7 A3
Rhiners Cl SO41 6 B4
Rhyme Hall Mews
SO45 10 D2
Ricardo Cres BH23 13 G4
Richards Cl SO45 10 B3
Richmond Cl SO40 26 B2
Richmond Ct SO41 21 D4
Ridgefield Gdns
BH23 14 B4
Ridgeway La SO41 18 D5
Ridgewood Cl SO45 16 B3
Rimbury Way BH23 12 A2
Ringbury SO41 19 E1
Ringwood By-Pass
BH24 25 A3
Ringwood Rd,
Ringwood BH24 8 A1
Ringwood Rd,
Southampton SO40 26 A5

Ringwood Rd,
Bransgore BH23 6 B1
Ringwood Rd,
Christchurch BH23 14 C1
Ringwood Rd,
Cadnam SO40 9 F3
River Ct SO41 19 F2
River Gdns SO41 21 E3
Riverdale Cl SP6 11 E2
Riverdale La BH23 12 A4
Riverlea Rd BH23 12 A4
Riversdale Rd BH6 12 A6
Riverside BH24 25 A4
Riverside Cl SO40 9 E4
Riverside Pk BH23 12 A5
Riverside Pl SP6 11 F3
Roberts Cl SO41 24 B5
Roberts Rd,
Hythe SO45 17 E2
Roberts Rd,
Southampton SO40 27 E6
Robin Gdns SO40 26 B3
Robin Gro BH25 23 A5
Robins Way BH23 13 G5
Rockbourne Gdns
BH25 15 G5
Rockery Cl SO45 16 C3
Rockleigh Dr SO40 4 D1
Rockram Cl SO40 9 D3
Rockram Gdns SO45 16 B3
Rodbourne Cl SO41 24 A5
Rodney Ct BH23 13 F4
Rodney Dr BH23 13 F4
Roebuck Cl BH25 22 D4
Roeshot Cres BH23 14 B3
Roeshot Hill BH23 13 H1
Roman Gdns SO45 16 C5
Roman Quay SP6 11 E4
Roman Rd SO45 16 C5
Roman Way SO45 16 C5
Romsey Rd,
Lynhurst SO43 20 B1
Romsey Rd,
Southampton SO40 9 B2
Rook Hill Rd BH23 13 H5
Rookcliff Way SO41 21 C3
Rookes La SO41 19 E5
Rooksbridge SO45 16 B3
Roscrea Cl BH6 12 B6
Roscrea Dr BH6 12 B6
Rose Cl SO45 17 F5
Rose Rd SO40 27 F5
Rosebery Av SO45 17 G5
Rosecrae Cl BH25 22 A3
Rosedale Cl BH23 13 E4
Rosehill Cl BH23 6 C1
Rosehill Dr BH23 6 B2
Roseleigh Dr SO40 26 D6
Rosemary Ct,
Southampton SO40 26 B4
Rosemary Ct,
Christchurch BH23 15 E5
Rosewood Gdns BH25 22 A3
Ross Rd BH24 25 E1
Rossiters Quay BH23 12 C4
Rossley Cl BH23 14 C3
Rothbury Cl SO40 26 C2
Rothbury Pk BH25 22 D4
Rotherfield Rd BH23 14 D3
Rothesay Dr BH23 14 C5
Rotterdam Dr BH23 12 C4
Round Copse SO45 16 B3
Round Hill SP6 11 E3
Roundhouse Dr SO40 26 B5
Rowan Cl,
Christchurch BH23 14 B4
Rowan Cl,
Lymington SO41 6 B6
Rowan Cl,
Southampton SO40 26 C6
Rowan Dr BH23 14 A4
Rowans Pk SO41 18 D4
Rowhill Dr SO45 16 B3
Royston Pl BH25 23 C7
Rubens Cl BH25 22 C3
Rufus Ct SO43 20 C4
Rufus Gdns SO40 26 B4
Rumbridge Gdns
SO40 27 F5
Rumbridge St SO40 27 F5
Rushford Warren BH23 13 F6
Rushington Av SO40 27 E5
Rushington La SO40 26 D6
Rushpole Ct SO45 16 B3
Russell Ct BH25 22 B4
Russell Dr BH23 12 D4
Russet Cl SO41 19 F4

Rye Dale SO40 4 B2
Rye Paddock La SO45 10 D1

Sadlers La SO45 17 F5
Saffron Dr BH23 13 H2
St Annes Gdns SO41 18 D4
St Davids Cl SO40 26 B2
St Denys*, Station Rd BH25 23 C5
St Francis Cl SO45 10 A6
St Francis Rd SO45 10 A6
St Georges Cl, Christchurch BH23 14 B5
St Georges Cl, Bransgore BH23 6 C2
St Georges Cres SP6 11 E2
St Georges Rd SP6 11 E2
St James Rd SO41 6 C5
St Johns St SO45 17 G2
St Margarets Av BH23 12 A5
St Marks Rd SO41 18 B4
St Mary Gro SO41 24 C3
St Marys Cl BH23 6 C2
St Michaels Cl SO45 10 A4
St Michaels Rd SO40 27 E3
St Thomas Pk SO41 19 E3
St Thomas's St SO41 19 E3
Salcombe Cres SO40 26 D5
Salcombe Rd SO40 26 D5
Salisbury Rd, Ringwood BH24 25 B1
Salisbury Rd, Ringwood BH24 25 B2
Salisbury Rd, Southampton SO40 26 C1
Salisbury Rd, Christchurch BH23 12 D1
Salisbury Rd, Fordingbridge SP6 11 F3
Salisbury Road Arc SO40 27 E4
Salisbury St SP6 11 F3
Salterns La SO45 10 C1
Samber Cl SO41 18 D2
Samphire Cl SO41 19 E1
Sanderlings BH24 25 E4
Sandilands Way SO45 17 G5
Sandle Copse SP6 11 B3
Sandleheath Rd SP6 11 A2
Sandlewood Cl SO40 26 C3
Sandmartin Cl BH25 23 A8
Sandown Rd BH23 13 E4
Sandy La SO43 20 B4
Sandy Plot BH23 12 C1
Saracen Cl SO41 18 D6
Sarah Sands Cl BH23 12 D2
Sarum Walk SO41 18 D1
Saulfland Dr BH23 14 A4
Saulfland Pl BH23 14 A4
Saxon Pl SO41 18 D1
Saxon Rd SO45 10 A5
Saxonford Rd BH23 14 A6
School La, Ringwood BH24 25 B3
School La, Milford on Sea SO41 21 E2
School La, Lymington SO41 19 E3
School Rd, Fawley SO45 10 C2
School Rd, Hythe SO45 17 F2
School Rd, Totton SO40 27 F5
Scotts Grn BH23 13 F2
Scotts Hills La BH23 12 D4
Sea Rd, Lymington SO41 21 E4
Sea Rd, New Milton BH25 15 H4
Sea View Rd BH25 15 G5
Sea Winds SO41 21 B3
Seacroft Av BH25 15 H5
Seafield Cl BH25 23 A7
Seafield Rd, Christchurch BH23 13 H4
Seafield Rd, New Milton BH25 15 H5
Seaton Cl, Lymington SO41 19 E1
Seaton Cl, Christchurch BH23 15 E4
Seaton Rd BH23 15 E4
Seaview Rd BH23 15 F3
Seaward Av BH25 15 H5
Seaward Rd SO45 17 G3
Seaway BH25 23 D7
Seaway Av BH23 13 H3
Second Marine Av BH25 23 B8
Second St SO45 10 B2
Sedgefield Cl SO40 26 C3
Selborne Rd SO40 26 D3
Sellwood Way BH25 15 G4
Serle Cl SO40 26 B4
Serle Gdns SO40 26 D5
Set Thorns Rd SO41 6 C5
Seventh St SO45 10 A1
Seymour Cl SO40 26 C2
Seymour Rd BH24 25 C1
Shackleton Sq BH23 6 C1
Shaftesbury St SP6 11 E4
Shaggs Mdw SO43 20 C4
Shakespeare Dr SO40 26 C2
Shamrock Way SO45 17 F1
Shappen Hill La BH24 8 B4
Sharpley Cl SP6 11 D2
Sharvells Rd SO41 21 C3
Shaves Cl BH25 22 C2
Shaw Cl SO40 26 C4
Shaw Rd BH24 25 E1
Shawford Cl SO40 26 A4
Shears Brook Cl BH23 6 C1
Sheldrake Gdns SO41 24 C2
Sheldrake Rd BH23 13 G5
Shelley Cl BH23 14 A5
Shelley Hill BH23 14 A5
Shelley Rd SO40 26 C2
Shelley Way SO41 21 E3
Shenstone Ct BH25 23 B5
Shepherd Cl BH23 14 C3
Shepherds Cl SO40 9 E4
Shepherds Hey Rd SO40 26 B2
Shepherds Rd SO40 9 E3
Sheridan Gdns SO40 26 C5
Sherringham Cl SO45 10 D2
Shetland Cl SO40 26 B3
Shinglebank Dr SO41 21 D3
Shires Walk SO40 26 B4
Shirley Cl BH23 6 C1
Shoblands Way SO45 17 G5
Shore Cl SO41 21 E4
Shore Rd SO45 17 G2
Shorefield Cres SO41 21 C2
Shorefield Rd SO41 21 A1
Shorefield Way SO41 21 C2
Shorts Cl BH23 12 C1
Shraveshill Cl SO40 26 C2
Shrubbs Av SO41 19 E2
Shrubbs Hill Gdns SO43 20 B4
Shrubbs Hilll Rd SO43 20 B4
Silkin Gdns SO40 26 C5
Silver St, Lymington SO41 24 A1
Silver St, Lynhurst SO43 20 A3
Silver St, Christchurch BH23 12 B4
Silver Way BH23 14 C5
Silverdale BH25 23 C7
Silvers End SO45 17 F5
Silvers Wood SO40 26 B2
Singleton Way SO40 26 B3
Sir Christopher Cl SO45 17 G2
Sizer Way SO45 16 B3
Sky End La SO41 24 C3
Slade Cl SO41 24 B2
Slades Hill SO45 10 B3
Slater Cl SO40 26 B3
Slinn Rd BH23 13 E3
Smith Cl SO45 10 B3
Smithy La BH25 22 A1
Smugglers La North BH23 14 B3
Smugglers La South BH23 14 B4
Smugglers Reach BH2 13 F5
Smugglers Wood Rd BH23 14 A3
Snellgrove Cl SO40 26 B1
Snooks La SO41 19 H1
Snowdrop Gdns BH23 13 H2
Solent Av SO41 19 G4
Solent Cl SO41 19 G4
Solent Ct SO41 21 C3
Solent Dr, New Milton BH25 23 B7
Solent Dr, Southampton SO45 17 E3
Solent Pines SO41 21 B3
Solent Rd, Southampton SO45 17 E6
Solent Rd, Christchurch BH23 15 F3
Solent Rd, New Milton BH25 15 G5
Solent Way SO41 21 F3
Somerford Av BH23 13 G2
Somerford Rd BH23 12 D4
Somerford Way BH23 13 E3
Somerley Vw BH24 25 B2
Somerton Cl BH25 22 E4
Somerville Ct*, Marryat Rd BH25 22 A4
Somerville Rd BH24 25 E2
Sopers La BH23 12 A5
Sopley Cl BH25 15 G5
Sopwith Cl BH23 13 G4
Sorrell Ct BH23 13 G2
Sorrell Way BH23 13 G2
South Av, Southampton SO45 10 A2
South Av, New Milton BH25 23 C5
South Baddesley Rd SO41 19 H3
South Gro SO41 19 F4
South Lawns Walk BH25 23 A6
South St, Lymington SO41 18 C4
South St, Southampton SO45 17 F4
South Sway La SO41 6 C6
South View Rd BH23 12 A4
South Weirs SO42 7 A4
Southampton Rd, Cadnam SO40 9 A/B2
Southampton Rd, Ringwood BH24 25 B3
Southampton Rd, Lynhurst SO43 20 C4
Southampton Rd, Lymington SO41 18 D1
Southampton Rd, Hythe SO45 16 D2
Southampton Rd, Fordingbridge SP6 11 F4
Southbourne Rd SO41 18 C4
Southcliffe Rd, New Milton BH25 15 G5
Southcliffe Rd, Christchurch BH23 13 H4
Southdown Ct*, Dorset Rd BH23 13 F2
Southern Gdns SO40 26 D4
Southern La BH25 23 A6
Southern Oaks BH25 23 A6
Southern Rd SO41 18 D3
Southey Rd BH23 13 E2
Southfield BH24 25 B4
Southfield La BH24 8 F3
Southfield Mews BH24 25 B4
Southlands SO41 18 C5
Southwood Av BH23 15 E3
Southwood Cl BH23 15 E3
Sovereign Cl SO40 26 B3
Spartina Dr SO41 19 E1
Speedwell Dr BH23 13 G2
Spencer Rd BH25 22 C4
Spicers Hill SO40 27 E6
Spicers Way SO40 27 E6
Spinacre BH25 23 C7
Spindlewood Cl BH25 23 B6
Spinney Dale SO45 17 G5
Spinney Gdns SO45 17 G5
Spinney Way BH25 22 C1
Spittlefields BH24 25 D3
Spring La BH25 23 E5
Spring Rd, Southampton SO45 17 G3
Spring Rd, Lymington SO41 19 G4
Springfield Cl SO41 19 G4
Springfield Dr SO40 26 D6
Springfield Gdns BH25 22 E4
Springwater Dr BH23 12 D4
Spruce Dr SO40 26 A3
Squirrels Walk SO45 17 E5
Stag Cl BH25 22 A3
Stallards La BH24 25 A3
Stanford Rise SO41 6 B5
Stanley Rd, Southampton SO40 26 D2
Stanley Rd, Christchurch BH23 15 E4
Stanley Rd, Lymington SO41 19 G5
Stannington Cl BH25 23 C5
Stannington Cres SO40 27 E3
Stannington Way SO40 27 F3
Stanpit BH23 12 D4
Staplecross La BH23 12 D2
Star La BH24 25 A3
Station Rd, Christchurch BH23 12 A3
Station Rd, New Milton BH25 22 B4
Station Rd, Lymington SO41 6 B5
Station Rd, Hinton BH23 14 C2
Station Rd, Fordingbridge SP6 11 C3
Station Rd, Ringwood BH24 8 F4
Station Rd, Southampton SO15 27 H4
Station Rd North SO40 27 G4
Station Rd South SO40 27 G4
Station St SO41 19 F3
Stella Ct BH23 15 E5
Stem La BH25 15 H2
Stephen Martin Gdns SP6 11 D3
Stephens Walk*, Kings Arms La BH24 25 A3
Stephenson Rd SO40 26 C1
Stibbs Way BH23 6 D1
Stirling Cl, New Milton BH25 22 D4
Stirling Cl, Southampton SO40 27 F3
Stirling Cres SO40 27 F3
Stirling Way BH23 13 G4
Stocklands SO40 26 B1
Stonechat Ct BH23 13 E3
Stonechat Dr SO40 26 B3
Stonehills SO45 10 D3
Stoneleigh Av SO41 24 A1
Stony La BH23 12 C1
Stony La South BH23 12 C4
Stopples La SO41 24 A1
Stour Rd BH23 12 A5
Stourbank Rd BH23 12 A4
Stouts La BH23 6 C2
Stratfield Pl BH25 22 A4
Stratford Pl SO41 18 D2
Strete Mount BH23 13 E3
Strides La BH24 25 A3
Strides Way SO40 26 B5
Stroud Gdns BH23 13 E4
Stroud La BH23 13 E4
Stroud Park Av BH23 13 E4
Stuart Rd BH23 15 E5
Stuckton Rd SP6 11 F4
Studland Dr SO41 21 C3
Studley Cl BH23 15 F4
Studley Ct BH25 15 G4
Summers La BH23 12 D1
Summertrees Ct BH25 22 E3
Sunderland Dr BH23 13 H3
Sundew Cl, New Milton BH25 22 F3
Sundew Cl, Christchurch BH23 13 H1
Sunningdale SO45 17 E4
Sunny Way SO40 27 E4
Sunnyfield Rd BH25 23 B7
Sunset Av SO40 27 E3
Sunset Rd SO40 27 E4
Surrey Cl SO40 26 D6
Sutton Pl SO42 7 D3
Sutton Rd SO40 26 D2
Swallow Cl SO40 26 B5
Swallow Dr SO41 21 F4
Swan Grn BH23 12 A5
Swan Mead BH24 25 E4
Sway Rd, Lymington SO41 18 A1
Sway Rd, New Milton BH25 22 B1
Sway Rd, Brockenhurst SO42 7 C6
Swordfish Dr BH23 13 G3
Sycamore Cl SO41 21 D2
Sycamore Rd, Lymington SO41 24 A1
Sycamore Rd, Southampton SO45 17 E
Sylvan Cl SO41 24 C
Sylvia Cres SO40 26 D

Talbot Dr BH23 14 D
Talbot Rd SO45 16 D
Tamar Gro SO45 16 D
Tamorisk Dr SO40 26 B
Tangmere Cl BH23 13 G
Tate Mdws*, Tate Rd SO40 27 H
Tate Rd SO15 27 H
Tates Rd SO45 17 G
Tattenham Rd SO42 7 C
Teal Cl SO40 26 C
Tebourba Cotts SO41 6 C
Tedder Way SO40 26 C
Tennyson Rd SO40 26 C
Tensing Rd BH23 12 D
Tern Cl SO45 17 H
Terrington Av BH23 14 B
Test La SO16 27 G1
Testbourne Av SO40 26 D
Testbourne Cl SO40 26 C
Testbourne Rd SO40 26 D
Testwood Av SO40 26 D
Testwood Cres SO40 26 C
Testwood La SO40 27 E
Testwood Pl SO40 27 F
The Barracks BH24 8 E
The Bartons SP6 11 E
The Boltons SO41 21 D
The Brackens SO45 16 D
The Bramblings SO40 26 E
The Brickyard SO40 9 D
The Bridges BH24 25 A
The Bucklers SO41 21 E
The Buttery BH23 12 D
The Cleeves SO40 26 C
The Cloisters, Lymington SO41 19 E
The Cloisters, Ringwood BH24 25
The Close, Ringwood BH24 25 E
The Close, Lymington SO41 6
The Close, New Milton BH25 23 C
The Coppice, Brockenhurst SO42 7 A
The Coppice, Christchurch BH23 13 C
The Crescent BH25 15 C
The Croft SO40 26
The Cross BH24 8
The Cul De Sac BH25 15
The Custards SO43 20
The Dell BH25 15
The Dorrits SO40 26
The Drive SO40 27
The Drove SO40 26
The Fairway BH25 23
The Fallows BH25 22
The Fowey SO45 10
The Furlong BH24 25
The Grange SO41 24
The Green SO41 21
The Hawthorns BH23 13
The Hyde BH25 22
The Lane SO45 10
The Lanes BH25 22
The Lawns BH23 15
The Linnets SO40 26
The Lydgate SO41 21
The Mall BH24 8
The Mallows BH25 22
The Marsh SO45 17
The Martells BH25 23
The Mead SO45 17
The Meadow BH25 15
The Meadows SO43 20
The Meadway BH23 14
The Mews*, Lepe Rd SO45 10
The Mount BH24 25
The Old Vineries SP6 11
The Orchard, Christchurch BH23 6
The Orchard, Lymington SO41 21
The Paddock, Southampton SO40 26
The Paddock, Brockenhurst SO42 7
The Paddocks SO45 10

he Pantiles SP6 11 C3
he Parade, Barton on Sea BH25 23 A6
he Parade, Southampton SO40 9 C2
he Parade, New Milton BH25 22 E4
he Park BH25 15 F5
he Pentagon SO45 10 B3
he Promenade SO45 17 G2
he Quadrant*, Brook Ter SP6 11 E4
he Quantocks SO45 16 D4
he Rampart SO41 19 E1
ne Redfords SO40 26 D3
ne Rise SO42 7 C3
ne Runway BH23 13 H3
ne Spinney SO40 26 B2
ne Square, Lymington SO41 18 C4
ne Square, Southampton SO45 10 D2
ne Sweep BH24 25 B3
ne Sylvans SO45 16 C4
ne Tins SO41 19 E3
ne Triangle*, Naish Est BH25 15 F5
ne Vale SO45 17 E4
ne Vinery BH25 22 D4
ne Wicket SO45 17 E4
ne Willows BH25 23 C7
ietchers Cl BH25 22 C2
iird St SO45 10 B1
iomas Cl SO40 26 C5
ioresby Ct BH25 15 H1
iornbury Av SO45 10 A5
ornbury Rd BH6 12 A6
orness Cl SO16 27 H2
ornfield Dr BH23 14 C3
ornham Rd BH25 22 E4
ornhill Cl SO45 10 A3
ornhill Rd SO45 10 A3
ree Acre Cl BH25 23 A6
ree Acre Dr BH25 23 A6
ursby Rd BH23 14 C3
demill Cl BH23 12 A2
fany Cl SO41 24 B1
burg Rd BH23 12 C4
ebarn La SO42 7 C6
nbermill Ct SP6 11 E4
nson Cl SO40 26 C5
he Barn SO41 19 E2
omer Cl SO45 10 B3
p La BH24 25 B3
rmead SO45 17 E4
land Ct SO41 21 C3
tton Western By-Pass SO40 26 A3
acey Ct BH23 14 D5
afalgar Ct BH23 13 F5
afalgar Pl SO41 19 F3
afalgar Way SO45 17 G5
anmere Cl SO41 19 F5
agore Rd SO40 26 C2
eebys Cl BH23 12 D1
eside BH23 14 A3
eside Av SO40 27 F5
foil Way BH23 13 H2
sillian Cl BH23 15 E3
sillian Way BH23 15 E3
vone Cl SO40 26 D6
nity Ct SO40 27 E1
ak Cl BH23 12 D2
cks Cl BH23 6 B2
dor Cl SO40 26 B3
rners Farm Cres SO41 24 C3
ynham Av BH23 12 A4
ers Cl SO41 19 E2
rells Ct BH23 6 C1
rells La BH24 8 B1

sdell Cl BH25 22 B3
dershore SO41 19 F1
dershore Rd SO41 19 F2
ands Av BH25 23 C7
per Common Rd SO41 18 A3
per Gordon Rd BH23 15 E4
Upper Mullins La SO45 17 E3

Vaggs La SO41 24 A1
Valiant Way BH23 13 G3
Valley Cl SO45 10 A4
Valley Rd SO40 27 E6
Valleydene SO45 17 F5
Vaughan Rd SO45 16 C3
Vecta Cl BH23 13 H4
Vectis Rd BH25 15 G5
Vellan Ct SO16 27 H2
Velsheda Ct SO45 17 F1
Velvet Lawn Rd BH25 22 A3
Verno La BH23 14 A3
Verwood Cres BH6 12 A6
Vetch Cl BH23 13 H2
Vian Ct BH25 22 B4
Vicarage Gdns SO41 24 B3
Vicarage La, Lymington SO41 24 B2
Vicarage La, Southampton SO40 9 D1
Vickery Way BH23 12 D3
Victoria Gdns, Fordingbridge SP6 11 C4
Victoria Gdns, Ringwood BH24 25 C4
Victoria Pl SO41 19 F4
Victoria Rd, Fordingbridge SP6 11 C3
Victoria Rd, Lymington SO41 21 C3
Victoria Rd, Christchurch BH23 13 E5
Viking Cl, Bournemouth BH6 12 A6
Viking Cl, Southampton SO45 10 B5
Viking Way, Bournemouth BH6 12 A6
Viking Way, Christchurch BH23 13 G6
Villette Cl BH23 12 A1
Villiers Rd SO45 17 E6
Vimoutiers Ct SP6 11 E3
Vincent Cl BH25 22 B4
Vincent Rd BH25 22 B4
Vinegar Hill SO41 21 D3
Viney Rd SO41 19 F6
Violet La BH25 22 C3
Viscount Dr BH23 13 G4
Vitre Gdns SO41 19 E5
Vixen Walk BH25 22 C2
Vulcan Way BH23 13 G3

Wagtail Dr BH25 22 A4
Wainsford Cl SO41 18 B4
Wainsford Ct SO41 18 C4
Wainsford Rd, Lymington SO41 18 A5
Wainsford Rd, Everton SO41 24 B5
Walhampton Hill SO41 19 F2
Walkers La North SO45 10 A4
Walkers La South SO45 10 A6
Walkford La BH25 15 G2
Walkford Rd BH23 15 F3
Walkford Way BH23 15 E3
Walnut Cl BH25 22 B4
Wanstead Cl BH24 25 D1
Warnes La BH24 8 B4
Warren Av BH23 13 F5
Warren Pk SO41 21 A1
Warren Pl SO40 26 C1
Warrys Cl SO45 17 F6
Warwick Av BH25 22 C4
Warwick Rd SO40 27 E3
Water La, Southampton SO40 26 C4
Water La, Hythe SO45 16 D5
Waterford Cl SO41 19 F4
Waterford Gdns BH23 14 D5
Waterford La SO41 19 F4
Waterford Pl BH23 15 E5
Waterford Rd, New Milton BH25 22 D4
Waterford Rd, Christchurch BH23 15 E5
Waterloo Rd SO41 19 F2
Waterloo Way BH24 25 C4
Watermans La SO45 17 E6
Watermead BH23 12 A5
Watermill Rd BH23 12 A2
Waters Green Ct SO42 7 D3
Waters Grn SO42 7 D2
Watership Dr BH24 25 E4
Waterside, Southampton SO45 17 F1
Waterside, Christchurch BH23 13 F6
Waterside Cl BH24 25 C1
Waterside Sq*, Waterside SO45 17 F1
Watery La BH23 13 F1
Watson Walk SO40 26 B5
Wavendon Av BH25 15 H4
Waverley Cl SP6 11 E2
Waverley Rd, New Milton BH25 23 C5
Waverley Rd, Fordingbridge SP6 11 E2
Wayside Cl SO41 21 D2
Wedgewood Gdns BH23 6 D1
Well Cl BH25 23 A5
Wellands Rd SO43 20 C3
Wellers Cl SO40 26 A4
Wellesley Av BH23 13 H3
Wellington Av BH23 14 A5
Wellington Cl SO45 16 D5
Wellington Ct BH25 22 C3
Wellington Pl SO40 9 C2
Wendover Cl BH25 23 A6
Wenlock Ct*, Hunt Rd BH23 13 F2
Wentwood Gdns BH25 23 E5
Wessex Av BH25 23 B5
Wessex Cl SO45 10 A5
Wessex Est BH24 25 D2
Wessex Rd BH24 25 D2
Wessey Cl BH23 13 H3
West Cl, Bournemouth BH6 12 A6
West Cl, Lymington SO41 18 B4
West Common SO45 10 A6
West Hayes SO41 19 F3
West Hill Dr SO45 17 E2
West La SO41 24 B4
West Rd, Lymington SO41 21 B2
West Rd, Christchurch BH23 6 B3
West Rd, Southampton SO45 16 D6
West St, Ringwood BH24 25 A3
West St, Southampton SO45 17 E2
West St, Fordingbridge SP6 11 E4
West View Rd BH23 12 D4
West Way SO41 18 C5
Westbeams Rd SO41 6 C5
Westbury Cl, Christchurch BH23 14 A3
Westbury Cl, Bransgore BH23 6 B3
Westbury Cl, New Milton BH25 23 C6
Westbury Rd BH24 25 D3
Westcliff Walk SO45 17 F1
Western Av BH25 15 H4
Western Rd SO41 18 D3
Westfield Gdns BH23 13 G1
Westfield Rd, Southampton SO40 27 E4
Westfield Rd, Lymington SO41 19 G4
Westgrove SP6 11 E3
Westlands BH23 6 B2
Westminster Ct BH25 23 A8
Westminster Rd SO41 21 B3
Westmoreland Ct SO41 24 A2
Westover Rd, Lymington SO41 21 D3
Westover Rd, Southampton SO16 27 H3
Westwood Ct SO40 27 E1
Westwood Rd SO43 20 B2
Wetherby Ct SO40 26 C3
Wetherby Gdns SO40 26 C3
Wharncliffe Ct BH23 14 D5
Wharncliffe Gdns BH23 14 D5
Wharncliffe Rd BH23 14 D5
Whartons Cl SO40 4 C2
Whartons La SO40 4 B2
Wheelers Walk SO45 10 A5
Whimbrel Ct BH23 13 F4
Whinfield Rd SO45 16 D6
Whitaker Cres SO41 18 C4
Whitby Ct SO41 21 C3
Whitby Rd SO41 21 B3
Whitcombe Cl SO40 26 D4
White Barn Cres SO41 24 C2
White Hart La SO40 9 B2
White Heather Ct SO45 17 F1
White Horses BH25 23 A8
White Knights BH25 23 B8
White Lion Courtyard BH24 25 A3
Whitecroft SO45 17 F4
Whitefield Rd BH25 22 B4
Whitehall BH23 12 B5
Whitehart Flds BH24 25 D2
Whitehayes Rd BH23 12 C1
Whitemoor Rd SO42 7 A3
Whites La SO45 10 D2
Whitewater Rise SO45 17 E5
Whitley Way BH25 22 C2
Whitsbury Rd SP6 11 D1
Whittington Cl SO45 17 E4
Wick Cl BH25 15 H2
Wick Dr BH25 23 A5
Wick La, Bournemouth BH6 12 A5
Wick La, Christchurch BH23 12 B5
Wickfield Av BH23 12 A5
Wickfield Cl BH23 12 B4
Wicklea Rd BH6 12 B6
Wickmeads Rd BH6 12 A6
Widbury Rd SO41 18 C5
Widden Cl SO41 6 C5
Wide Lane Cl SO42 7 C3
Widecombe Dr SO45 16 D3
Wildburn Cl SO40 26 B1
Wilde Cl SO40 26 C5
Wildfell Cl BH23 12 A1
Wildground La SO45 17 F6
William Cl BH23 15 E2
William Ct BH23 15 E5
William Rd SO41 19 E2
Willow Av SP6 11 E2
Willow Dr BH23 12 A5
Willow Pl BH23 12 B4
Willow Walk BH25 23 C8
Willow Way BH23 12 A5
Willowdene Cl BH25 22 D4
Wilton Gdns BH25 22 A4
Wiltshire Gdns BH23 6 A2
Wiltshire Rd BH23 6 A2
Wilverley Cl SO41 18 C5
Wilverley Pl SO45 10 A4
Wilverley Rd, Brockenhurst SO42 7 C3
Wilverley Rd, Christchurch BH23 13 F3
Winchester Rd BH25 22 D2
Winchester Way SO40 26 B5
Windermere Gdns SO40 27 E4
Windmill Cl SO41 21 D3
Windmill Copse SO45 17 E5
Windrush Way SO45 17 F4
Windsor Cl SO41 24 A1
Windsor Ct BH23 15 E5
Wingate Rd SO40 26 D4
Wingfield Av BH23 14 B3
Wingrove Rd SO40 4 B2
Winsford Cl BH23 14 A3
Winsor La SO40 9 F1
Winsor Rd, Cadnam SO40 9 D3
Winsor Rd, Southampton SO40 27 F5
Winston Ct BH25 23 B6
Winston Ct*, Stour Rd BH23 12 A3
Winston Way BH24 25 C2
Winton Cl SO41 19 G4
Winton Way BH25 22 C2
Wisbech Way SO41 24 B3
Wolfe Cl BH23 12 D3
Wolsey Way SO41 21 E3
Wood La, Brockenhurst SO42 5 A2
Wood La, Lymington SO41 21 D3
Wood Lawn Cl BH25 23 A6
Wood Lodge SO40 26 B2
Wood Rd SO40 4 B3
Woodcock La SO41 24 C2
Woodfield Gdns BH23 14 B3
Woodford Cl BH24 25 D3
Woodgreen Walk SO40 26 B3
Woodhayes Av BH23 14 B3
Woodland Way, Christchurch BH23 14 B5
Woodland Way, Lymington SO41 21 C3
Woodlands Cl, Christchurch BH23 6 B2
Woodlands Cl, Southampton SO45 17 E6
Woodlands Ct SO45 17 E6
Woodlands Rd, Southampton SO40 4 A4
Woodlands Rd, New Milton BH25 23 A8
Woodlands Rd, Brockenhurst SO42 7 C5
Woodlands Way BH25 22 C1
Woodley Gdns SO41 18 D2
Woodruff Cl BH23 13 H2
Woodside Av SO41 19 F5
Woodside Cl SO41 19 F5
Woodside Gdns SO40 4 B3
Woodside La SO41 19 E5
Woodstock La BH24 25 B3
Woodvale Gdns BH25 22 D3
Woodville Rd SO45 10 C2
Worcester Pl SO41 19 F4
Worthy Rd BH25 22 B4
Wortley Rd BH23 14 D5
Wren Cl, New Milton BH25 23 A5
Wren Cl, Ringwood BH24 25 E4
Wren Cl, Christchurch BH23 13 G5
Wychwood Dr SO45 10 A6
Wykeham Pl SO41 19 E4
Wyndham Cl BH23 15 F3
Wyndham Rd BH23 15 F2

Yaldhurst La SO41 18 B3
Yarrell Mead SO41 18 B3
Yarrow Cl BH23 13 H2
Yelverton Av SO45 17 F6
Yeoman Pk SO16 27 H2
Yeomans Way SO40 27 E4
Yeovilton Cl SO41 24 B5
Yerville Gdns SO41 24 A2
Yew La BH25 22 D3
York Av BH25 22 C4
York Cl SO40 27 F3
York Pl BH25 22 C4

Edition 386 E 07.01